GLOBAL WARMING, AND CLIMATE CHANGES
Transparency and Accountability

GLOBAL WARMING, AND CLIMATE CHANGES

Transparency and Accountability

(In 3 Volumes)

Transparency and Accountability of Global Environment

(Volume-3)

Editor

Gopal Bhargawa

ISHA BOOKS

DELHI-110 033

Global Warming and Climate Changes
Transparency and Accountability

Rs. 2500 (Set)
Rs. 835 (each volume)

ISBN: 81-8205-078-2 (Set)
ISBN: 81-8205-081-2 (Vol. 3)

Published in 2004 in India by
Isha Books, D-43, Prithviraj Road,
Adarsh Nagar, Delhi-110 033

Laser Typesetting at : Shree Vaishnavi Creations, Delhi
Printed at : Chawla Offset Press, Delhi

Contents

Preface

The global environment is in a constant state of flux because of natural causes and human modifications for food production, settlements, infrastructure, or to produce and trade goods. Most intentional changes are designed to harness the environment for human benefit. Domestication of land for intensive food production is one example; harnessing river resources to provide fresh water, energy and transport is another. Such changes may also unintentionally alter the quality or quantity of environmental resources and be difficult to cope with.

Analysing old and new threats to human security shows that human vulnerability to environmental conditions has social, economic and ecological dimensions. The most conspicuous and widely reported manifestation of this vulnerability is when people are affected suddenly and violently by natural hazards. These events turn into disasters when local communities are not able to cope with their impacts.

Although everyone is vulnerable to environmental impacts of some kind, the ability of people and societies to adapt to and cope with change is very varied. Developing countries, particularly the least developed, have less capacity to adapt to change and are more vulnerable to environmental threats and global change, just as they are more vulnerable to other stresses. The coping capacity of human society is a combination of all the natural and social characteristics and resources available in a particular location that are used to reduce the impacts of hazards. These include factors such as wealth, technology, education, information, skills, infrastructure, access to resources and management capabilities.

The present volume provides authoritative and valuable information on various issues concerning global environment. Student, researchers and teachers, besides environmental scientists and environment activists will find this work informative and useful.

Editor

1

Global Environment: An Outlook

Human activities contributing to land degradation include unsuitable agricultural land use, poor soil and water management practices, deforestation, removal of natural vegetation, frequent use of heavy machinery, overgrazing, improper crop rotation and poor irrigation practices. The 1992 Earth Summit took a step forward in focusing attention on problems associated with land resources.

Deforestation over the past 30 years has been the continuation of a process with a long history. By the time of the Stockholm Conference, much forest cover had already been removed. Major direct causes of forest clearance and degradation include expansion of agricultural land, overharvesting of industrial wood, fuelwood and other forest products, and overgrazing. Underlying drivers include poverty, population growth, markets and trade in forest products, as well as macroeconomic policies.

Forests are also damaged by natural factors such as insect pests, diseases, fire and extreme climatic events. The net loss in global forest area during the 1990s was about

94 million ha. This was the combined effect of a deforestation rate of 14.6 million ha annually and a rate of reforestation of 5.2 million ha annually. Deforestation of tropical forests is almost 1 per cent annually. In the 1990s, almost 70 per cent of deforested areas were changed to agricultural land, predominantly under permanent rather than shifting systems.

A recent study using globally comprehensive and consistent satellite data estimated that the extent of the world's remaining closed natural forests (where crown cover is more than 40 per cent) in 1995 was 2 870 million ha, about 21.4 per cent of the land area of the world. The Stockholm Conference recognized forests as the largest, most complex and self-perpetuating of all ecosystems, and emphasized the need for sound land and forest use policies, ongoing monitoring of the state of the world's forests and the introduction of forest management planning.

Today, the Stockholm Conference recommendations relating to forests remain valid and unfulfilled, in many ways, because of conflicting interests in managing forests for environmental conservation and economic development. Global biodiversity is being lost at a rate many times higher than that of natural extinction due to land conversion, climate change, pollution, unsustainable harvesting of natural resources and the introduction of exotic species.

Land conversion is most intensive in tropical forests and less intensive in temperate, boreal and arctic regions; atmospheric nitrogen deposition is largest in northern temperate areas close to cities; introduction of exotic species is related to patterns of human activity. Human population growth together with unsustainable patterns of consumption, increasing production of waste and pollutants, urban development and international conflict are further contributory factors to biodiversity loss.

Over the past three decades, decline and extinction of species have emerged as major environmental issues. Although insufficient information is available to determine precisely how many species have become extinct in the past three decades, about 24 per cent of mammals and 12 per cent of bird species are currently regarded as globally threatened. The past three decades have been marked by the emergence of a concerted response to the biodiversity crisis.

Civil society, including a hugely diverse and increasingly sophisticated NGO network, has been a major driving force behind this. Increased stakeholder participation relating to conservation action has seen the emergence of partnerships between NGOs, governments and the private sector. A number of international conventions have been developed that deal specifically with conservation of threatened species. These include the 1973 Convention on International Trade in Endangered Species of Wild Fauna and Flora (CITES) and the 1979 Convention on the Conservation of Migratory Species of Wild Animals (CMS). A major policy response of the 1990s is the adoption, ratification and implementation of the Convention on Biological Diversity (CBD).

About one-third of the world's population lives in countries suffering from moderate-to-high water stress - where water consumption is more than 10 per cent of renewable freshwater resources. Some 80 countries, constituting 40 per cent of the world's population, were suffering from serious water shortages by the mid-1990s. Increasing water demand has been caused by population growth, industrial development and the expansion of irrigated agriculture.

For many of the world's poorer populations, one of the greatest environmental threats to health remains the continued use of untreated water. While the percentage of people served with improved water supplies increased from 79 per cent (4.1 billion) in 1990 to 82 per cent (4.9 billion) in 2000, 1.1 billion people still lack access to safe drinking water and 2.4 billion lack access to adequate sanitation. Most of these people are in Africa and Asia. Lack of access to safe water supply and sanitation results in hundreds of millions of cases of water-related diseases, and more than 5 million deaths, every year.

Large, but poorly quantified adverse impacts on economic productivity have been noted in many developing countries. Emphasis on water supply, coupled with weak enforcement of regulations, has limited the effectiveness of water resource management, particularly in developing regions. Policy makers have now shifted from supply to demand management, highlighting the importance of using a combination of measures to ensure adequate supplies of water for different sectors. Measures include improving water use efficiency, pricing policies and privatization.

There is also a new emphasis on integrated water resources management (IWRM), which takes into account all the different stakeholders in water resource planning, development and management. Marine and coastal degradation is caused by increasing pressure on both terrestrial and marine natural resources, and on the use of the oceans to deposit wastes. Population growth and increasing urbanization, industrialization and tourism in coastal areas are root causes of this increased pressure.

In 1994, an estimated 37 per cent of the global population lived within 60 km of the coast - more people than inhabited the planet in 1950. The effects of population are multiplied by both poverty and human consumption patterns. Globally, sewage remains the largest source of contamination, by volume, of the marine and coastal environment, and coastal sewage discharges have increased dramatically in the past three decades. Marine and coastal eutrophication from elevated nitrogen inputs has emerged as a worrying trend not foreseen three decades ago. There is increasing evidence that blooms of toxic or otherwise undesirable phytoplankton are increasing in frequency, intensity and geographic distribution.

Severe eutrophication has occurred in several enclosed or semi-enclosed seas, including the Black Sea. Human-induced changes in the natural flow of sediment have emerged since the Stockholm Conference as a major threat to coastal habitats. Urban and industrial development drives the construction of residential and industrial infrastructure which, depending on its nature, can alter sediment flow. There is particular concern about the possible effects of global warming on coral reefs.

During the intense El Niño of 1997-98, extensive coral bleaching occurred on coral reefs worldwide. While some reefs quickly recovered, others, particularly in the Indian Ocean, Southeast Asia, the far western Pacific and the Caribbean, suffered significant mortality, in some cases more than 90 per cent. Progress in protecting the marine and coastal environment over the past 30 years has generally been confined to relatively few, mostly developed countries, and to a relatively few environmental issues. Overall, coastal and marine environmental degradation not only continues but has intensified.

Acid precipitation has been one of the most prominent environmental concerns over the past decades, especially in Europe and North America, and more recently also in China. Thousands of lakes in Scandinavia lost fish populations due to acidification from the 1950s to the 1980s. Significant damage to forests in Europe became a high priority environmental issue around 1980. Air pollutant emissions have declined or stabilized in most industrialized countries, largely as a result of abatement policies developed and implemented since the 1970s. Initially, governments tried to apply direct control instruments but these were not always costeffective.

In the 1980s, policies were directed more towards pollution abatement mechanisms that relied on a compromise between the cost of environmental protection measures and economic growth. Stricter environmental regulation in industrialized countries has triggered the introduction of cleaner technology and technological improvements, especially in the power generation and transport sectors. Since the industrial revolution, the concentration of CO_2, one of the major greenhouse gases, in the atmosphere has increased significantly, contributing to the greenhouse effect known as 'global warming'.

The increase is largely due to anthropogenic emissions of CO_2 from fossil fuel combustion and to a lesser extent land-use change, cement production and biomass combustion. Greenhouse gas emissions are unevenly distributed between countries and regions. Organization for Economic Cooperation and Development (OECD) countries contributed more than half of CO_2 emissions in 1998, with a per capita emission of about three times the world average. However, the OECD's share of global CO_2 emissions has decreased by 11 per cent since 1973.

Climate change represents an important additional stress on those ecosystems already affected by increasing resource demands, unsustainable management practices and pollution. The United Nations Framework Convention on Climate Change and the Kyoto Protocol are the key policy instruments adopted by the international community to try to address the problem of greenhouse gases emissions. The protection of the ozone layer has presented one of the major challenges over the past 30 years, spanning the fields of environment, trade, industry, international cooperation and sustainable development.

The depletion of the ozone layer has now reached record levels, especially in the Antarctic and recently also in the Arctic. In September 2000, the Antarctic ozone hole covered more than 28 million km^3. Continuous efforts by the international community have resulted in a marked decrease in the consumption of ozone-depleting substances. The ozone layer is predicted to start recovering in the next one or two decades and to return to pre-1980 levels by the middle of the 21st century- if the control measures of the protocols to the Vienna Convention are adhered to by all countries.

About half of the world's population (47 per cent) now lives in urban areas, compared to little more than onethird in 1972. The accumulation of people, their consumption patterns, travel behaviour and their urban economic activities impact the environment in terms of resource consumption and waste discharges. Some 70 per cent of the world's urban population live in Africa, Asia or Latin America. The urban population is expected to grow by 2 per cent per year during 2000-15, and to reach an overall 65 per cent by 2050.

The implications of rapid urban growth include increasing unemployment and poverty, inadequate urban services, overburdening of existing infrastructure, lack of access to land, finance and adequate shelter, and environmental degradation. Managing the urban environment sustainably will therefore become one of the major challenges for the future. Poverty is among the major drivers of urban environmental degradation. The urban poor, who are unable to compete for scarce resources or protect themselves from harmful environmental conditions, are most affected by the negative impacts of urbanization. It is estimated that one-quarter of the urban population lives below the poverty line and that female-headed households are disproportionately affected.

Inadequate waste collection and waste management systems are the cause of serious urban pollution and health hazards, especially in cities in developing countries. Cities in industrialized countries also face the consequences of past environmentally damaging production techniques and inadequate waste disposal. Well-planned, densely populated settlements can reduce the need for land conversion, provide opportunities for energy savings and make recycling more cost-effective.

People and the environment are suffering increasingly from the effects of natural disasters due to high population growth and density, migration and unplanned urbanization, environmental degradation and possibly global climate change. The number of people affected by disasters rose from an average of 147 million a year in the 1980s to 211 million a year in the 1990s. While the number of geophysical disasters has remained fairly steady, the number of hydrometeorological disasters (such as droughts, windstorms and floods) has increased.

In the 1990s, more than 90 per cent of those killed in natural disasters lost their lives in hydrometeorological events. While floods accounted for more than twothirds of people affected by natural disasters, they are less deadly than many other types of disaster, accounting for only 15 per cent of deaths. The most expensive disasters in purely economic terms are floods, earthquakes and windstorms but events such as drought and famine can be more devastating in human terms.

While earthquakes accounted for 30 per cent of estimated damage, they caused just 9 per cent of all fatalities due to natural disasters. In contrast, famine killed 42 per cent but accounted for just 4 per cent of economic damage over the past decade. Among the least developed countries, 24 of the 49 face high levels of disaster risk; at least six of them have been affected by between two and eight major disasters per year in the past 15 years, with long-term consequences for human development. Since 1991, more than half of all the disasters reported occurred in countries with medium levels of human development. However, two-thirds of those killed came from countries with low levels of human development, while just 2 per cent came from highly developed countries.

Some experts link the recent trend in extreme weather events to an increase of the global mean temperature. Many parts of the world have suffered major heat waves, floods, droughts and other extreme weather events. A number of major accidents involving chemicals and radioactive materials have drawn attention worldwide to the dangers of mismanagement, particularly in the transport, chemical and nuclear power sectors. These events often have impacts that transcend national boundaries; they also emphasize the fact that issues of technological safety concern more than just the developed countries.

Biodiversity will continue under threat if there is no strenuous policy action to curb human activity. Continued urban and infrastructure expansion, plus the increased impacts of climate change, severely deplete biodiversity in most regions in all scenarios. Pressures will also increase on coastal ecosystems in most regions and scenarios.

Growing populations and increased economic activity, particularly in agriculture, will lead to increased demand for freshwater in most scenarios. Similarly, the demands for food and the ability to meet them in the different scenarios reflect a combination of shifts in supply and demand, influenced by social, economic and environmental policies. In Markets First, even with a decrease in the percentage of the population facing hunger, the total number affected changes relatively little and even increases in some regions as populations grow.

First the targeting of hunger reduction as a key goal, and the emphasis on more balanced development between regions, help to achieve dramatic reductions in the percentages and total numbers of people affected. The sharp increases in most regions in Security First points to the unsustainability of such a scenario in terms of social acceptability.

In Africa, there is increasing risk of land degradation. In Policy First and Sustainability First, easier access to support services helps farmers to manage soils better and policies based on integrated land management become commonplace in the region. At the other end of the spectrum, in a Security First scenario, while reasonable conditions are maintained in the protected areas serving the land-owning elite, the high concentration of people elsewhere contribute to severe land degradation and soil erosion. Similar problems arise in Markets First as better quality agricultural land is taken over for commodity and cash crop production.

Under the Markets First scenario in Asia and the Pacific, water withdrawals are expected to increase in all sectors, leading to an expansion of areas with severe water stress in South and Southeast Asia. Slower economic growth under Security First tempers growth in demand. With effective policies and lifestyle changes under the Policy First and Sustainability First scenarios, water withdrawals remain at current levels or even decrease in most of the region.

The ability of Europe to address the issues of large-scale air pollution and greenhouse gas emissions will depend heavily upon developments in the areas of energy use and transportation. Extremely active policies to improve public transportation and energy efficiency can be expected in Policy First and Sustainability First worlds, but not in Security First or even Markets First circumstances.

Land and forest degradation as well as forest fragmentation remain among the most relevant environmental issues in Latin America and the Caribbean in all scenarios. Significant loss of forest area occurs in a Markets First scenario. In a Security First world, the control over forest resources by transnational companies that create cartels in association with the national groups in power, promotes the growth of some forest areas, but this is not enough to stop net deforestation. More effective management ameliorates some of these problems in Policy First. Unsound deforestation stops almost completely in a world of Sustainability First.

As the world's biggest emitter of greenhouse gases, North America plays a major role in determining the future climate of the planet. In Markets First, the region's refusal to participate significantly hampers international efforts to control the emissions of these gases, and per capita and absolute emissions remain high. Under Policy First, emissions are reduced through increased fuel efficiency and greater use of public transport but most spectacular results are achieved in Sustainability First.

West Asia is one of the most water-stressed regions of the world, with more than 70 million people living in areas under severe water stress. Under the Markets First and Security First scenarios, population and economic growth lead to strong increases in withdrawals for households and industry, resulting in an increase in areas with severe water stress and affecting over 200 million people by 2032.

A range of policy initiatives help to counteract additional demands related to economic growth in both Policy First and Sustainability First. Although total withdrawals drop in both scenarios, water scarcity persists and demand continues to exceed available water resources. Fish and other marine stocks are a key area of concern in the polar regions. Under Markets First, massive increase in commercial harvesting and abandonment of targeted fisheries leads to some fish populations crashing.

Illegal, unregulated and unreported fishing activities cease in Security First under direct pressure from powerful regulatory interests, but controlled exploitation rises to very high levels. Total collapse of any single fishery is averted under Policy First by enforcement of stringent harvesting quotas and other regulatory systems. In Sustainability First, fish and marine mammals are rigorously defended against overexploitation. The environmental implications of the various scenarios illustrate the legacy of past decades and the level of effort that will be needed to reverse powerful trends.

One of the major policy lessons from the scenarios is that there can be significant delays between changes in human behaviour, including policy choices, and their environmental impacts, specifically:

— Much of the environmental change that will occur over the next 30 years has already been set in motion by past and current actions.

— Many of the effects of environmentally relevant policies put into place over the next 30 years will not be apparent until long afterwards.

Issues in Environmental Reporting

Over the years, the pollution of air, water and soil and their causes, impacts and interlinkages, have been further specified as to single out policy-relevant environmental issues. These are now both widely recognized and can be addressed by policy makers and researchers in quantitative terms. Climate change, ozone depletion and urban air pollution are well-known examples, but the list usually includes a dozen or so other environmental problems.

From all this, the most prominent problems related to the environment and common to most world regions, can be singled out as being:

— Land use & agriculture (incl. nature, biodiversity)

— Energy & climate change

— Urban stress & health

From these three issues, problems such as climate change, acidification, eutrophication, water stress, soil degradation, biodiversity loss and ozone depletion have been singled out in several assessment studies. Obviously, disaggregating the environmental system implies that interlinkages and cycles need to be addressed because of dependencies and interactions among the '(sub)themes' e.g. acidification and climate change, biodiversity and land use / forests. In fact, these interlinkages are highly policy relevant since they could point to synergy among the measures taken.

Agricultural development, increasing energy use and urbanization, linked by trade and transport, are the main processes and activities affecting natural ecosystems and human health. Society reacts through the formulation of policies, laws, agreements and conventions, and the application of accompanying instruments to reach the targets, usually by either reducing pressures (emission reductions), and/or limiting impacts (end-of-pipe measures). In practice, these are highly complicated processes.

Over the last few years, there have also been several attempts to define information needs for assessment of sustainable development and human security, including the World Bank's Genuine Savings Rate, UNCSD's Sustainable Development Framework, IHDP's Human Security Index and UNDP's Human Development Index and Human Poverty Index. But work on underlying core data sets as follow-up to the Bangkok Symposium and IEA/GEO data initiatives was very limited, probably because (i) the task goes beyond everybody's remit and not one organization has the mandate to take full responsibility and (ii) the ever returning financial constraints meant that even more modest proposals to improve the data situation were never put into practice.

Substantial regional differences can be seen in the contribution to this aspect of the climate change problem. The USA, Russia and China are clearly the three biggest producers of CO_2 in the world. While searching for policy-relevant causes of environmental degradation, an attempt was made to analyse the problem of deforestation. The major deforestation took place in tropical forest areas of Africa, South America, western Asia and in some parts of the Asia-Pacific region.

In order to explore socio-economic factors responsible for deforestation, the total roundwood production (million m3 per year) was looked at. It turns out that several main producers of roundwood in the world (US, Brazil, India, China, Russia and Canada) are not characterized by high values of change in forest area (i.e. with the exception of China, Brazil).

Besides roundwood production, another important factor for deforestation, which shows the use of 'traditional' fuel as a percentage of total fuel use. The consumption of traditional fuels such as firewood and charcoal occurs mainly in Africa, South and South East Asia, the Mekong Basin and South America, notably in the Caribbean. Somewhat surprisingly, traditional fuel use was fairly high in certain parts of Central Europe, particularly in areas affected recently by conflicts i.e. Albania and the states of former Yugoslavia.

Deforestation occurs mainly in Africa and South and Southeast Asia - the result of growing pressure of the population on forests for the use of traditional fuel. But such a conclusion may be too simple. Wood is a very profitable commodity and in many countries a large part of roundwood production is illegal and not captured in official statistics. Usually population size is used to distribute economic activities over grid cells. This holds both for financial measurements (GDP) and physical processes such as production and consumption of fossil fuels. This spatial distribution mechanism needs to be checked against available information on physical parameters within the economy. Land use, a socio-economic variable, can be derived by combining country statistics collected by FAO at the global level with physical information on soils, climate and land cover.

Such land use data sets are the result of processing and integration of 'raw' data collected through physical measurement, surveys and country collections, including Olson's World Ecosystems, FAO Soil Map, and FAOSTAT. From land use, indicators for terrestrial biodiversity can be derived as was done for GEO-1 i.e. domesticated vs. natural area. Land cover is usually obtained through remote sensing.

At the global level, mainly NOAA satellite imagery is currently used; at regional levels Landsat, SPOT or radar imagery are also used. Land cover, together with land use, is used for initial distribution of land as a starting point for projections of global terrestrial changes in the future. Land cover is also used to estimate the degree of fragmentation of natural areas as one of the pressures on biodiversity. The recent 1x1 km data set needs further checking against available regional and country data such as EEA/CORINE Land Cover and RIVM/Staring Centre European Land Use data sets.

Historic land use changes are to be improved to better enable trend analysis and model calibration. Global soil maps, produced by FAO, need constant checking and updating. Gridded climate data are derived by interpolation of station measurements, mainly undertaken by IIASA/RIVM and CRU on the basis of WMO inputs, for different periods in time and with somewhat different methodologies. There is still little information on the composition and characteristics of ecosystems and biogeographical areas.

Distribution to gridcells is done through population size in the absence of economic information for small, subnational units. The results of this 'top-down' method need to be compared with the national reported 'bottom-up' emission levels, which are now starting to appear. Information on the state of the environment needs to be translated to exposure of humans and ecosystems to agents detrimental to their health and quality, respectively.

For health impacts, this can be done by combining demographic data with dose-effect relationships for specific pollutants, resulting in impacts on the status of human health in terms of mortality, morbidity, and disabilities. At the global level, the main bottleneck is the poor demographic and health data at the national and subnational levels.

Degradation of ecosystems can be expressed by the vulnerability to environmental pollution. Highly vulnerable ecosystems will degrade easily with an increase in sulphur of nitrogen, while others can absorb heavy loads. The exceedance of critical loads can be used as an indicator of potential impacts of degradation and pollution.

Furthermore, potential land productivity can be compared to actual yields as determined by prevailing agricultural practices. Decreasing potential productivity or yields are a sign of a shift to unsustainable use of natural resources. In addition, new methodologies are required to express the dimensions of biodiversity (habitat area and ecological quality). The habitat area approach is currently under development through analyses of combined pressures on ecosystems. Such a conditional approach needs further work and integration with biodiversity quality, combined with new data collection activities.

To a large extent, core data sets are being used by many other organizations involved with assessment, such as UNEP, FAO and UNDP, UNCSD and OECD. Certainly the input data on the D-P side of the D-P-S-I-R chain (economy, population, land use, energy) are common to many assessment activities. Underlying observation (surveys, measurements, remote sensing) form the most basic data layer, with considerable overlap among several global assessments. Emissions and concentrations still overlap substantially, while going further to the other side, damages and losses are more specific to each individual assessment.

Examples of Environmental Issues

Ozone depletion

One of the effects of ozone depletion is an increase in the incidence of skin cancer. The destruction of ozone results from increased emissions of CFCs and other halocarbons into the atmosphere over the last few decades. This, again, has resulted from the production and consumption of ozone-depleting substances (ODS) contained in air conditioners, sprays, refrigerators and other goods. Although production and consumption of ODS are successfully being reduced, impacts such as skin cancer incidence will increase in the coming decades due to the delay in recovery of the ozone layer.

Climate change may hamper this through stratospheric cooling, which facilitates the destruction of ozone. The risks are derived from a dosetime response model (UV-Chain) for the relation between skin cancer induction and change in UV. Maximum values are estimated for around 500 extra cases over the normal 1100 per million people per year in North West Europe. Assuming a population of 160 million in North West Europe, this accounts for some 80,000 extra people every year.

The risk is related to biological skin characteristics and age distribution and cannot be extrapolated to other regions without more information on human sensitivity to UV and demographic patterns. Higher skin cancer incidence results from increases in the UV dose at ground level. The carcinogenic UV dose is ascertained by weighting the ambient UV spectrum according to the carcinogenic effectiveness. Increases have been most profound in Western Europe, although the highest risks are found more southwards.

The increase in UV results from depletion of the ozone layer in the atmosphere. Photochemical breakdown releases active chlorine and bromine atoms, which catalyze ozone destruction. The decrease of ozone concentration can be derived from satellite imagery such as TOMS and Nimbus data. In the case of CFCs, emissions occur by and large when these substances are consumed or applied ; the production itself is a minor source. Sales are often used as a proxy for consumption.

Increasing industrial production, transport, agriculture and other economic activities result in higher emission levels of acidifying compounds, mainly sulphur and nitrogen. Acidification occurs when terrestrial ecosystems can no longer absorb the deposition of SO2, NOx and NH3 without showing damage to soils and vegetation.

When critical levels have been exceeded, the health of the ecosystem is at risk. Acidification is being studied for Europe within the framework of the UN Convention on Long-Range Transboundary Air Pollution. Data on emissions and air transport by EMEP and on the sensitivity of ecosystems by CCE is integrated by IIASA in the RAINS model. RAINS-optimization scenarios directly support the negotiations under the convention on future emission reduction obligations by the countries in Eastern and Western Europe. Similar RAINS calculations are now also being made for Southeast Asia, within the RAINS-Asia project sponsored by the World Bank.

Eutrophication

Industrial, agricultural and other anthropogenetic activities have modified global bio-geochemical nitrogen (N) cycles by doubling the natural rate of nitrogen fixation. This is reflected in increasing emissions of nitrogen gases into the atmosphere, increasing riverine transport of nitrogen into the oceanic environment, accumulation in inland aquatic systems and deposition to natural ecosystems. Because nitrogen is the primary nutrient limiting biological production in many terrestrial and freshwater and marine environments, increases in nitrogen inputs (in any form) can alter those ecosystems through eutrophication.

The main nitrogen substances emitted by human activities are nitric oxide (NO) and nitrogen dioxide (NO2) (together denoted as NOx), and ammonia (NH3). Global emissions of NOx and NH3 amount to 56 and 54 million ton N per year, respectively. Livestock production is by far the dominant source for atmospheric NH3, while industrial sources and fossil fuel combustion are the most important sources of NOx. Regionally, this pattern varies, for example, in regions with extensive rural areas or where large-scale burning is in progress.

The calculated 1992 deposition onto global land areas amounts to and 4.4x1012 eq N per year. The highest mean regional nitrogen deposition rates of 1200 mg nitrogen per m2 per year occur in Eastern Europe. South Asia and OECD Europe have mean deposition rates of 80-90 meq nitrogen per m2 per year, respectively. Other regions with somewhat lower deposition rates are Japan and East Asia. The highest maximum deposition rates occur in East and South Asia and somewhat lower deposition rates occur in Western and Eastern Europe, North America, South East Asia, Western and Eastern Africa and Japan.

Scattered regions of high susceptibility occur in South America and Africa. In tropical forests the low soil pH is the major cause of low critical load values, while in other parts the ecosystem itself is highly susceptible, locally exacerbated by low soil pH. Intermediate susceptibility occurs in the western USA, Europe and Russia. Deposition exceeds critical loads for eutrophication in 7-18% of the global area of (semi-) natural ecosystems. Large parts are affected by nitrogen deposition, particularly in Europe and North America.

In the former USSR 9% of the (semi-)natural ecosystems is affected mainly in remote tundra and taiga ecosystems. Within the group of developing countries, Asian and African countries have extensive natural and semi-natural areas where critical loads are exceeded. The results suggest that, apart from the heavily industrialised regions, a number of regions with low population densities, such as South America and Africa, and remote regions in Canada and the former USSR, may become or already be affected by nitrogen eutrophication.

Scenarios indicate that agricultural nitrogen (fertilizer and animal waste), the dominant source of NH3, may double between 1995 and 2025 in tropical countries. Scenarios for anthropogenic NOx emissions excluding agriculture indicate an increase of 50%, with no or only slight increase in North America and Europe. Major increases come from southern and eastern Asia, the Middle East, Africa and Latin America. This suggests that in many tropical regions nitrogen emissions and deposition may sharply increase in the near future, and that problems of nitrogen-related eutrophication may be aggravated and expand over larger areas in these regions.

Greenhouse effect

The increasing atmospheric concentrations of greenhouse gases (GHGs: CO_2, CH_4, N_2O, CFCs and others) will lead to a change in radiative forcing and, hence, to changes in global and regional climate. Emissions of GHGs stem from several natural processes and ecosystems, and human activities. Emissions are increasingly dominated by human activities, such as the burning of fossil fuels for energy and deforestation. Central to the DPSIR chain of climate change are the global bio-geochemical cycles of carbon and nitrogen. These cycles determine the final atmospheric concentrations through a broad range of processes and their interactions.

For example, plants absorb CO_2 through photosynthesis, which ecosystems can either store or release. The rate of most processes is dependent on locally prevailing climatic conditions. The resulting change in climate has impacts on many natural and human systems. For example, the potential to produce food and other resources may change, sea levels may rise, distribution of plant and animal species may shift, and the frequency and magnitude of storms and other extreme weather events may change.

A decade ago, the first Integrated Assessment Models (IAM) for the climate system emerged. Initially, these models were straightforward energy-balance models, which were coupled to emission-energy models. Fluxes from land-use change (i.e. deforestation) were generally prescribed. Environmental impact models were sometimes attached as well. These early IAMs were still strongly natural science-based, but showed that without climate policies, emissions would continue to increase, leading to a doubling of the concentration of greenhouse gases by the middle of next century.

Gradually, the IAMs were further developed. The first broad integration of the DPSIR chain in climate-change models has been achieved in IMAGE-2 and TARGETS. These models include more advanced routines that determine changes in energy use and land use, and include technology improvements and changes in demand and resource availability. They further incorporate a wide range of impacts on sea level, ecosystems and agriculture, and also quantify and calculate many important feedback mechanisms within the climate system.

The TARGETS model, moreover, emphasizes different responses by explcitly including several cultural perspectives. The models, currently being integrated to develop a single, broad and consistent integrative global environmental assessment tool, enable projections and analysis of environmental change over the next century. The environmental and societal data sets needed to initialize these models are obtained from international organizations (e.g. World Bank, OECD and FAO) and research programmes (e.g. IGBP and IHDP).

Scenarios with largely different assumptions on demographic, economic and technological developments, and with or without dedicated climate policies, have been developed and used to support negotiations with the Framework Convention on Climate Change (FCCC) and the scientific assessments of IPCC and IGBP. The scenario simulations illustrate the major negative consequences that will occur if no action is taken through climate policies. These include:

— A doubling of current atmospheric levels of CO_2 by 2050 and important negative impacts on crops, natural vegetation and sea levels in all regions;

— The rate of increase of impacts on world vegetation, which could be larger in the first half of the next century than ever before or thereafter.

— Impacts on vegetation (agricultural and natural) which will slow down in the second half of the next century, but sea levels that will continue to rise substantially.

Cross-Cutting Requirements

DPSIR framework to environmental issues also reveals

several broad data issues deserving special attention. These relate to the geographic scale and spatial resolution of assessment, to the need to encompass policy responses and to a broadening of the concept of environment to environmental sustainability. These cross-cutting issues are addressed here, again, by pragmatic and recent examples.

Water assessment

Theoretically, there is no water problem at the global scale. Not even one tenth of the amount that falls on land is currently consumed by society. Even when looking at the continental scale, there are hardly shortages of water. The water table of Africa is positive in the sense that availability is higher than actual use, and this situation will continue for a long time. But going further down the scale to the natural unit for water resources, i.e. river basins, problem areas can be easily noted, not only in the daily lives of farmers and city dwellers, but also among societies and countries, as witnessed by numerous overt or latent conflicts.

Many relevant issues actually occur on even smaller units in both rural areas (irrigation) and urban areas (shortage, pollution). From the assessment point of view, there is growing awareness that discrepancies between the demand and the availability of fresh water pose a threat to sustainable development. In other words, development of society and economy is in competition with ecosystem integrity. Both water quantity and water quality are at stake here.

The situation seems to be getting worse in many areas all over the world. Predicted global changes may aggravate water stress in the future, in particular, climate changes.

The water issue requires research and assessment at these different scales in order to serve as instruments in solving the problem:

— *Global*: assessment of the nature and magnitude of water problems in order to prepare for proper policy actions at the international level (UNEP, CSD); some quality aspects are strongly connected with global issues; e.g. the nutrient impairment among regions stems from the uneven distribution of demand for and supply of food.

— *Regional:* assessment of the consequences of large-scale economic and social developments for the aquatic environment, and assessment of the impairment of demand and supply, both in time and place. Possible solutions must be obtained in regional agreements and conventions (e.g. EU); water-quality aspects sometimes have a regional component. Nevertheless, regionally differentiated approaches still need a common approach.

— *Catchment:* assessment of management options to support actual management in river basins; a comprehensive approach is especially challenging for international rivers; management at this level has to be agreed upon in treaties. At this level a comprehensive approach incorporating quality and quantity aspects will be necessary. Guidance and supportive methods provided by international bodies like UNEP can strengthen such assessments substantially.

In fact, it seems desirable to go even further down the scale i.e. to the local level of many relevant activities in society such as irrigation and urbanization. For instance, it would be somewhat misleading to conclude that there are hardly any water problems in the Niger River basin, while urban and rural dwellers experience frequent shortages in many parts of the tributary river basins.

A cell-based approach, which allows the integration of complex activities and processes (urbanization, irrigation), seems more appropriate. In addition , the issue of time scale and horizon could be added here: an area might experience problems once every 10 years on average; but in practice it may happen four years in a row. Impacts, such as those from climate change, may only be apparent on long-term time scales.

In support of global water assessments, various institutes (e.g. SEI, WRI, RIVM) have already developed databases, models and other tools. Applications have been made at various levels of detail and for various regions, but often in isolation. Further international cooperation and harmonization could benefit the assessment work substantially. In any case, a critical challenge in freshwater assessments is how to deal with different geographical scales.

Information on driving forces and socio-economic developments is usually available at the country level. Further refinement to areas of, for example, 50 by 50 km is needed for several variables, such as irrigation, population density and land use. Climate changes are relevant at the global and continental level. The actual freshwater assessments are only meaningful when based on calculations at the catchment level and below. Translations back and forth among these levels and processes require the use of dedicated geographic information systems (GIS).

The comprehensive assessment of the fresh water resources by the UN-DPCSD compared present and future water withdrawal and availability on a country-by-country basis. This already provides insight into a number of the problems to be expected with water supply for personal consumption and irrigation. For various regions, however, the full extent of water problems is not recognized because of the scale of assessment. For the world as a whole, water stress was analyzed for GEO-1 using calculations at the catchment level.

Presently, global water problems are more thoroughly assessed at the catchment and even subcatchment level, using clustering techniques to recognize similarities between problems in different regions and taking into account the local scale of relevant processes. With global climate change as a distinct possibility, different regions of the world will have to cope with a different water situation than currently is the case. Nitrogen (next to phosphorus) is the primary nutrient limiting plant, algae and microbial production in many freshwater and marine environments.

Increases in nitrogen inputs (in any form) can alter those ecosystems. For example, shifts in plant and algal species composition, increased productivity, decreased species diversity, changes in foodweb structure, as well as development of hypoxic (low oxygen) and anoxic (no oxygen) conditions in aquatic systems (with consequences for fish and other higher organisms) can result from increased nitrogen inputs. World-wide, about 58% of the total export of Dissolved Inorganic Nitrogen (DIN) of 20.8 Mt nitrogen per year stems from fertilizers, while human sewage and deposition account for 24 and 18%, respectively.

The contributions to nitrogen export from fertilizer use, deposition and human population show very similar patterns, with the highest nitrogen export in OECD Europe, Eastern Europe and southern and eastern Asia, largely reflecting the distribution of agricultural production intensity and human population. Given human-induced acceleration of nitrogen fixation and other changes in the cycle, it is no surprise that nitrogen concentrations in surface waters have increased over time.

In freshwater systems with sufficient phosphorus, addition of inorganic nitrogen can cause eutrophication. This can occur either independently or coupled to acidification. Decreased diversity of both animal and plant species generally accompanies both eutrophication and acidification. Most often, eutrophication in estuaries and coastal seas is caused by anthropogenic nitrogen loading, in sharp contrast with freshwater systems in which phosphorus is the element limiting net primary production and controlling eutrophication.

Eutrophication can cause anoxia and hypoxia, both of which appear to be becoming more prevalent in many estuaries and coastal seas. Conditions of low oxygen concentration have resulted in significant losses of fish and shellfish resources. Increasing anoxia has been observed in the Baltic Sea, Black Sea and Chesapeake Bay, while hypoxic events have increased in the North Sea, the Kattegat and Long Island Sound. Eutrophication is also associated with a loss of diversity, both in the benthic community and among planktonic organisms, as manifested by the incidence of nuisance algal blooms in many estuaries and coastal seas.

For aquatic ecosystems, it is difficult to establish a system of critical loads for nitrogen. This is caused by the complexity of the interactions between nitrogen and phosphorus in the eutrophication of freshwater systems, and the high seasonal and interannual variability of weather conditions causing runoff and associated nitrogen inputs and water discharge. Therefore, for aquatic ecosystems the anthropogenic nitrogen export is used as an indication of the potential for eutrophication.

Experiences over the last decades have shown that several important relationships exist between global and local environmental problems , and other issues such as human and economic development, population growth, poverty and changing political structures. In fact, many environmental problems arise largely from fundamental social or economic problems and cannot be solved by isolated measures in relation to environmental technology and legislation.

The concept of 'sustainable development' has been put forward as a guiding concept for policy making to design development policies that appropriately address the economic and social needs and aspirations of a growing global population, while maintaining the quality of the environment and its resource base. It will become increasingly important to clearly place information on the environment within a sustainable development context.

The first Global Environment Outlook addressed mainly environmental issues as such. GEO-1 indicated the relative importance of these issues at the regional level and presented current trends. Environmental concerns appear to change over time, depending on the transitional stage of society. Human and ecosystems' health and global environmental issues are usually not the highest priority in mainly agriculture-based economies.

Environmental concerns related to urbanization and the use of natural resources often become more important in the transition towards industrialized and wealthy service-oriented economies. The environmental policy responses to these problems change accordingly from sectoral policies to market-based initiatives. Assessment for sustainable development requires a broadening of both the assessment and policy horizon. It requires a comprehensive approach to all the economic, social and ecological issues involved. This is a difficult task, in part because sustainable development is a concept linked to values that depend on one's cultural perspective. If one considers nature to be fragile, a precautionary type of environmental policy seems to be appropriate. However, if one considers nature to be robust, a virtually unlimited exchange of natural capital for economic and social capital is appropriate.

In the end society, and therefore politicians, will have to decide what is sustainable and what is not. The uneven distribution of problems and opportunities in time and space will have to be taken into account when these decisions are made. Therefore, these decision-makers must be provided with the best scientific data and information in easily understandable form. Sustainable development simultaneously includes social, economic, environmental and institutional objectives.

Consequently, assessment for sustainable development needs to address these four domains and their interaction. During the past few years several organizations and institutes have focused on assessment for sustainable development, in particular, through the use of indicators. An interesting step was made by the World Bank, which proposed to focus on different types of capital. Indicators of sustainable development will have to reflect the size of economic, ecological and social/human capital and the degree to which people actually have access to such capital.

The potential use of indicators of sustainable development can be illustrated to a certain extent by some indicators currently being developed or already accepted among the international community. These include indicators for economic, social and natural capital (World Bank), human development (UNDP) and biodiversity pressures (Convention on Biodiversity, RIVM). The savings rate could indicate the vitality of an economy because it illustrates the capability to invest in new developments.

Latin America has also experienced a negative genuine savings rate, but there the trend has been reversed to a positive trend. With this method, the genuine savings rates in 'rich' regions are currently positive. The costs of solving environmental problems are - in comparison with national income - relatively low in these regions. The social dimension of sustainability can be illustrated by the Human Poverty Index (HPI), as developed by UNDP. This index represents sets of variables expressing the percentage of people expected to die before the age of 40, access to social resources and access to natural resources.

The number of people living in poverty -following the definition of HPI - is likely to continue to decrease until 2010. However, the numbers in Africa and India will still increase. Only a minority of the African population is not facing poverty at all. Access to natural resources will become a more important cause of poverty in developing countries. Access to adequate food and clean water is also an important factor in the improvement of public health. Another major environmental health problem is air pollution, both in the developed and in developing countries.

Air pollution by Suspended Particulate Matter (SPM) causes about 6% of the 50 million deaths that occur globally per year. Due to growing fossil fuel combustion, as well as increases in traffic and industrial emissions, air pollution is expected to become even more important, especially in countries in transition and in developing countries. The area of land used for human purposes in an extensive or intensive manner may increase by about 20 % in the period 1970-2050. Assuming a Business-as-Usual scenario, it is expected that in the next century some 40 % of the total land will have been domesticated, leaving around 35 % for forest, grassland or savannah.

The most rapid decrease in natural land takes place in Africa and Asia. The pace of conversion of land is already slowing down in North America and Europe; it might even be reversed. Agricultural reform is the main cause of these changes in land use patterns. World-wide, however, land conversion from nature to agriculture will be the most prominent cause of loss of natural capital. In addition, physical factors such as salinization, erosion and lowering of watertables, and pollution may have detrimental effects on the quality of ecosystems.

2

State of the Environment

There have been some impressive gains in human development, particularly in the developing world: incomes and income poverty have improved, people are living longer, are healthier, more literate and better educated than ever before. Average annual incomes in developing countries have mostly risen: in real terms they rose during 1972-99 by 13 per cent in Africa, by 72 per cent in Asia and the Pacific and by 35 per cent in Latin America and the Caribbean; in West Asia they fell by 6 per cent. Nevertheless, the challenges remain daunting in the 21st century, with high levels of deprivation persisting across the world.

Approximately 1.2 billion people, or one-fifth of the world's population, still live in extreme poverty on less than US$1 a day, and 2.8 billion people, or almost half the world's population, on less than US$2 a day. Three-quarters of those in extreme poverty live in rural areas, and the majority are women. Poverty is not limited to developing countries: more than 130 million people in the developed countries of the Organization for Economic

Cooperation and Development (OECD) are considered income-poor.

Ill health is related to environmental factors and poverty. Medical innovation, progress in basic health care and enabling social policies have resulted in dramatic increases in life expectancy and declines in infant mortality. Overall, a child born today can expect to live eight years longer than one born 30 years ago. However, poverty rates in both urban and rural areas, as well as major infectious diseases such as HIV/AIDS, tuberculosis and malaria are a threat to health successes over the past few decades.

Tremendous progress has also been made in terms of education in the past 30 years, and adult literacy rates have increased from an estimated 63 per cent in 1970 to 79 per cent in 1998. Nevertheless, in 2000 there were still 854 million illiterate adults, of whom 543 million were women, and 325 million children not attending school of whom 56 per cent were girls. Improved education together with capacity building is considered to have been critical in slowing down world population growth from an annual peak of 2.1 per cent in the early 1970s to 1.3 per cent in 2000. World population increased from about 3.85 billion people in 1972 to 6.1 billion in mid-2000, and is currently growing by 77 million people a year.

Most of the growth is concentrated in developing regions, with nearly twothirds in Asia and the Pacific. Six countries accounted for 50 per cent of the annual growth: India, China, Pakistan, Nigeria, Bangladesh and Indonesia. The population of industrialized regions, currently 1.2 billion, is expected to change little in the next 50 years while that of the developing regions is projected to increase from 4.9 billion in 2000 to 8.1 billion by 2050. This difference is mainly due to levels of fertility.

Less developed countries recorded fertility at 3.1 children per woman over the period 1995-2000, while developed countries recorded fertility at 1.57 children per woman over the same period - well below the replacement level of 2.1 children per woman. The gap in life expectancy between the developing and the more developed regions narrowed from 22 years in 1960 to less than 12 in 2000. Between 1995 and 2000, life expectancy in the industrialized regions was estimated to be 75 years, compared to about 63 years in developing regions.

As world fertility continues to decline and life expectancy rises, the population of the world will age faster in the next 50 years than during the past half-century. Nevertheless, the emergence of HIV/AIDS in the 1970s has depressed life expectancy in the most affected countries, and the disease is now the fourth most common cause of death worldwide. More than 60 million people have been infected with HIV since the 1970s, and 20 million have died. Of the 40 million living with HIV/AIDS worldwide, 70 per cent are in sub-Saharan Africa where it is the leading cause of death.

Compared to Africa, the relative impact of the HIV/AIDS epidemic in most other regions of the world is still low. There have been notable population movements over the past 30 years due to rapid urbanization, international migration and the movement of refugees and displaced persons. The number of people living outside their own countries rose from 84 million in 1975 to an estimated 150 million people at the end of the 20th century. The number of refugees rose from 2.7 million in 1972 to a peak of 18.3 million in 1992, and stood at 11.7 million at the end of 1999.

By 2001, an estimated 22 million people 'of concern' were numbered among the world's refugees, returnees and persons displaced within their own countries. The more developed regions are expected to continue being net receivers of international migrants, with an average gain of about two million a year over the next 50 years. Because of low fertility in recipient countries, this migration will have a significant impact on population growth in the developed regions.

Development of Economy

The world economy has, on aggregate, expanded considerably over the past three decades, despite significant fluctuations. World gross national product (GNP) more than doubled from approximately US$14 300 billion in 1970 to an estimated US$29 995 billion in 1999. However, these figures do not include the value of environmental goods and services which are critical to the Earth's life-support systems, and contribute to human welfare but are outside the market. An estimate of the economic value of these ecosystem services is between US$16 000 billion and US$54 000 billion a year, with an average of US$33 000 billion a year. This estimate should be considered a minimum because of the nature of the uncertainties.

The world economy grew by 3.1 per cent annually in real gross domestic product (GDP) between 1980 and 1990, and 2.5 per cent annually between 1990 and 1998, with annual per capita growth rates of 1.4 and 1.1per cent respectively. There has, however, been significant regional variability over this period, with by far the highest growth rates in Asia and the Pacific, which contains more than half the world population. Per capita GDP almost doubled in Northwest Pacific and East Asia during 1972-99, growing by an annual average of 2.4 per cent a year; by contrast, it fell in sub-Saharan Africa.

Despite global economic growth, the gap between rich and poor has widened both between developed and developing countries and within countries, particularly in Latin America and sub-Saharan Africa. Per capita incomes have risen only marginally in most regions, with the exceptions of Europe and North America. Currently 3.5 billion people in low-income countries earn less than 20 per cent of the world's income, while the 1 billion people living in developed countries earn 60 per cent. The ratio between income earned in countries with the richest 20 per cent of the population, compared to the world's poorest 20 per cent, has also widened - from 30:1 in 1960, to 60:1 in 1990, to 74:1 in 1997.

Growth in use of energy and transport are both indicators of economic development, and both have severe impacts on the environment. Private vehicular transportation has become an entrenched lifestyle choice among those who can afford it. Since the 1970s, about 16 million new vehicles have come onto the world's roads annually and passenger cars account for 15 per cent of total global energy consumption. Inequalities in income are also reflected in similar disparities in material consumption. It has been estimated that the richest 20 per cent of the world's population accounts for 86 per cent of total private consumption expenditure, consumes 58 per cent of the world's energy, 45 per cent of all meat and fish, 84 per cent of paper, and owns 87 per cent of cars and 74 per cent of telephones.

Conversely, the poorest 20 per cent of the world's population consumes 5 per cent or less of each of these goods and services. For many developing countries poverty, unemployment and low productivity are major concerns. In developing countries as a whole, the informal sector provides 37 per cent of employment, and as much as 45 per cent in Africa. In the 1980s, structural adjustment programmes (SAPs) were introduced by the World Bank to correct underlying economic imbalances and improve economic efficiency through reforms.

SAPs have had economic, social and environmental impacts, including negative impacts on social stability and environmental sustainability. Poverty, unemployment and falling standards of living also emerged as significant problems for countries in economic transition in the 1990s. One critical issue is that of external debt which stood at US$2 572 614 million in 1999. The Heavily Indebted Poor Countries (HIPCs) initiative was launched in 1996 and by November 2001 debt-reduction packages totalling US$36 000 million had been committed to 24 countries. However, there has been some disappointment with the initiative, and many of the countries receiving HIPC debt relief still spend more on debt servicing than on basic education or health.

Growth of Science and Technology

Science and technology have brought about major breakthroughs over the past 30 years in, for example, the fields of information and communications, medicine, nutrition, agriculture, economic development and biotechnology. Forty-six global hubs of technological innovation have been identified around the world, principally in Europe and North America. Information and communications technology (ICT) particularly has revolutionized the way people live, learn, work and interact.

The Internet, mobile phones and satellite networks have shrunk time and space. Satellite communications technology from the mid-1980s gave rise to a powerful new medium with a global reach. Bringing together computers and communications in the early 1990s unleashed an explosion of ways to communicate, process and store, and distribute enormous amounts of information. In 2001, more information could be sent over a single cable in a second than was sent over the entire Internet for a month in 1997.

ICT is advancing rapidly, presenting tremendous opportunities for human development by making it easier for more people to access available information from remote locations, quickly and cheaply. However, the uneven diffusion of ICT means that access to related technological developments may be an advantage for the minority only. Today, Internet users are predominantly urban and 79 per cent of users live in OECD countries, which contain only 14 per cent of the world's population.

Nevertheless, even in developing countries the increase in Internet use has been dramatic - for example, from 3.9 million to 33 million people in China between 1998 and 2002. Mobile telephones have overcome the infrastructure constraints of fixed lines and the number of subscribers has increased from slightly more than 10 million around the world at the start of the 1990s to more than 725 million at the beginning of 2001, or one mobile phone for every eight inhabitants.

Additionally, new technologies are helping people to better understand the environment. In July 1972, the US government launched the first LANDSAT satellite. By 2002, the LANDSAT programme has acquired 30 years of records which constitute the longest continuous record of data on the Earth's continental surfaces. This has added a new dimension to environmental monitoring and assessment, enabling changes to be tracked, trends monitored and early warning improved. However, for some developing countries, technology can be a source of exclusion instead of a tool for progress.

'Technology is created in response to market pressures, not the needs of poor people who have little purchasing power. As a result research neglects opportunities to develop technology for poor people'. For example, of the 1,223 new drugs marketed worldwide between 1975 and 1996, only 13 were developed to treat tropical diseases. New technologies also come with unforeseen risks to human health and the environment: for example, ozone-layer depletion from the use of CFCs, the side effects of drugs, unintended use of new technologies as weapons, pollution, concerns over the impacts of genetically modified organisms, and technological disasters such as Chernobyl and Bhopal.

Governance: An Overriding Issue

Governance is an overriding issue which applies at all levels and sectors of society - from local to global, from the private to the public sector - and has impacts on law and human rights, political, parliamentary, democratic and electoral systems, civil society, peace and security, public administration, public information, the media and the corporate world. Awareness of and attention to governance issues have consequently grown in every aspect of modern life, not least in relation to the environment.

'Good' governance is recognized as a prerequisite for sound policy development and as being even more important for ensuring that policies are effectively implemented. The 30 years between the Stockholm Conference and the 2002 World Summit on Sustainable Development have witnessed a major overhaul of political systems, including the decolonization of Africa, the end of apartheid in South Africa, the raising of the Iron Curtain, the reunification of Germany and the disintegration of the former Soviet Union.

There has been a rapid increase in economic and trade alliances, and in membership of the United Nations, which stood at 190 in March 2002. Perhaps the greatest change has been at the national level where democracy and transparency have become major issues, particularly since the 1990s, and governments have been replaced as a result of popular demand. In the past 10 years, more than 100 developing and transition countries ended military or one-party rule.

The internationalization of the environment since 1972 has had a bearing on national politics in many of the developed countries, with green parties making their mark, particularly during the 1980s. The past three decades have also seen the massive growth of civil society movements in all the regions of the world, with many organizations taking a more proactive role. Nongovernmental organizations are increasingly influencing and sometimes participating in government and private sector decisions.

The power of the people in influencing policy has been recognized since at least the 1970s: 'It is only through the deep concern, information and knowledge, commitment and action of the people of the world that environmental problems can be answered. Laws and institutions are not enough. The will of the people must be powerful enough, insistent enough, to bring about the truly good life for all mankind.' The growing integration of finance, economies, culture, technologies and governance through globalization is having profound impacts, both positive and negative, on all aspects of people's life and the environment.

With market forces beginning to dominate the social and political as well as economic spheres, there is a danger that power and wealth become concentrated in a minority while the majority become increasingly marginalized and dependent. In the 21st century, the challenge is to institute stronger governance to ensure that globalization operates for the benefit of most people and not just for profits. The African region has a wealth of natural resources, offering myriad opportunities for human, social and economic development.

Its diversity of cultures and valuable indigenous knowledge provides the necessary human capital to realize these opportunities. Nevertheless, Africa entered the 21st century facing enormous challenges. Population (millions) by sub-region: Africa Human development Of the 49 countries in Africa for which data are available, 20 are classified as having medium human development and 29 low human development. Sub-regional differences are most marked between Northern Africa, where rapid progress has been made over the past 30 years, and sub-Saharan Africa.

Income distribution is also highly skewed with the poorest 10 per cent of Africans earning less than 5 per cent of the income and the richest 10 per cent earning 25-45 per cent of the income, depending on the country. There are also large discrepancies between the status of men and women in Africa, with women prohibited from owning property or land in many societies. Women also generally earn less and do not hold a representative proportion of high-ranking jobs.

Access to health services varies but generally lags behind the international average. Poor economic growth and increasing population pressures on existing facilities have contributed to low investment in the health sector. In 1998, government expenditure on health care per capita ranged from a high of US$623 in South Africa to only US$15 in Madagascar. Africa has had one of the highest population growth rates in the world over the past 30 years; at the current 2.4 per cent a year, it is much higher than the global average of 1.3 per cent. The population more than doubled from 375 million in 1972 to 794 million in 2000, or approximately 13 per cent of the world's population.

The HIV/AIDS pandemic, which killed 2.3 million people in Africa in 2001, is having an impact on all aspects of human, social and economic development. The continent has the highest new infection rate and the largest proportion of the population living with HIV/AIDS in the world. In 2001, there were 28.1 million people with HIV/AIDS in sub-Saharan Africa - 70 per cent of the global total. Over the past 20 years, the disease has had a severe impact on life expectancy in the region, and in countries such as Botswana and Malawi average life expectancy is already below 40 years.

Northern African countries are less severely affected. Despite commendable action by many countries, the impact of AIDS on social development, economic growth and on health systems is projected to run to billions of dollars. For example, in South Africa the impact is anticipated to be equal to 0.4 per cent of GDP over the next decade. The impact on families, communities and societies is incalculable. The economies of African countries have been largely based on primary products or extraction of natural resources, both exported unprocessed. As a result, economic growth has been below potential, because revenues from the value that is added by processing accrue outside the continent, making African economies extremely vulnerable to external price fluctuations and trade regulations.

The first oil crisis, in 1973-74, sparked a series of setbacks and economic recessions that have lasted for more than a quarter of a century. Falling prices for coffee, cocoa and other cash crops during the 1980s had catastrophic impacts on the economies of the region. Between 1970 and 1995, Africa lost half its markets, representing a loss of income of about US$70 billion a year. Africa's dependence on rain-fed agriculture means that production is vulnerable to climatic variability, which can severely affect food and human security, and exports.

The focus on mineral extraction, cash crops and timber harvesting has also had detrimental impacts on the environment. With the additional constraints of a growing population, Africa's economic performance has been poor over the past 25 years. Annual growth of per capita GDP for sub-Saharan Africa was -1 per cent between 1975 and 1999, and incomes have fallen. Nevertheless, 34 African countries recorded increases in per capita incomes between 1994 and 1997, and 18 grew at aggregate rates above the 5 per cent a year threshold for reducing poverty. There is some speculation that this may signal a sustained economic recovery, partly reflecting the positive results of implementing growth-oriented macroeconomic and structural reforms.

Since the mid-1990s price controls have largely been lifted, marketing boards abolished, trade taxes rationalized, financial markets liberalized and the process of privatization accelerated. National external debt is still a significant barrier to economic growth and poverty reduction in Africa. For the region as a whole, it increased almost 22-fold from US$16 960 million in 1971 to US$370 727 million by 1999. In 1970, the debt burden of sub-Saharan Africa was just US$6 000 million, or 11 per cent of GNP; this grew to US$330 000 million or 61 per cent of GNP by 1999. Since then there has been a small decline.

In Northern Africa, growth of external debt has followed a similar pattern. Recently, more emphasis has been placed on debt relief and increasing foreign direct investment (FDI). Although 20 African countries have had debt-reduction packages approved under the enhanced Heavily-Indebted Poor Countries Initiative, indebtedness remains an issue of major concern.

In terms of technological achievement, diffusion of, access to and adoption of science and technology is generally poor in Africa. The region has a number of marginalized countries (such as Ghana, Kenya, Mozambique, Senegal, Sudan and Tanzania) where large parts of the population have not even benefited from the use of old technology such as fixed-line telephones. Nevertheless, Africa has a number of 'dynamic adopters' in the use of new technology, including Algeria, Egypt, South Africa, Tunisia and Zimbabwe, although the diffusion of old technology is still slow and incomplete even in these countries. The region has two recognized global hubs of technological innovation: in El Ghazala, Tunisia, and Gauteng, South Africa.

However, most African countries are not included on the Technology Achievement Index (TAI), and even South Africa, which has the highest score of any African country, has a TAI of 0.340 which is less than half the highest TAI value of 0.744 for Finland. Lack of access to affordable appropriate technologies seriously constrains sustainable development options. In the agricultural sector, for example, many African countries depend on irrigation but the more efficient drip irrigation systems are too costly for most farmers, resulting in wastage of water.

Africa also remains marginalized with respect to access to biotechnology for agricultural productivity, pharmaceutical products and disease prevention. Foreign companies may be able to exploit biological species commercially whereas local companies do not have the technology, capital or know-how to do so. All countries in the region have Internet connections with a total of around 4 million users, or one user for every 200 people, compared to a world average of about one user for every 30 people.

Mobile communications technology has enabled Africa to leapfrog the infrastructure constraints of land-based communications. In 1990, African countries had no or low access to cellular communications; by 1999 the technology had spread dramatically to most countries - for example, availability has risen from 0.1 per 1 000 population to 132 per 1 000 in South Africa.

Despite significant institutional and political changes over the past 30 years, 'good' governance in Africa is still a partial and fragile feature. Most notable changes have been the end of colonialism, as well as military and one-party rule in most countries, and the rise in participation by non-governmental, community-based and civil society organizations. Nevertheless, corruption is endemic in many countries. For example, 14 African countries scored less than five out of a clean score of ten on a corruption perception index, and four of those scored less than two.

A further barrier to stability, economic growth and social development in the region over the past 30 years has been the escalating incidence of civil conflict. Conflicts, typically arising from ethnic or religious differences or competition for natural resources, have resulted in massive displacement of people and diversion of financial resources away from vital sectors such as health and education. About one in every five Africans still lives in circumstances of civil conflict.

The Asia and the Pacific region occupies 23 per cent of the Earth's land area and is home to more than 58 per cent of its people. The region includes several of the world's largest emerging economies, such as China and those of Southeast Asia. Over the past 30 years, the region has gradually moved from a subsistence lifestyle towards a consumer society, with rapid rates of urbanization and westernization as well as population increase. This transition has not been without adverse social and economic, as well as environmental, impacts.

Of the 53 countries in the region, 7 are classified as having a high level of human development, 21 as a medium level, and 5 as a low level (the other countries, mainly small islands in the South Pacific, have not been classified). The Northwest Pacific and East Asia sub-region has made sustained and rapid progress in most areas of human development, while South Asia lags behind with human and income poverty still high. Per capita GDP (in US$1995) ranges from US$506 in South Asia to US$4 794 in Northwest Pacific and East Asia.

Adult literacy rates show similar improvements increasing from 33 to 55 per cent in South Asia during 1972-99, and from 55 to 84 per cent in Northwest Pacific and East Asia. An estimated three-quarters of the world's poor live in Asia, and poverty is particularly significant in Afghanistan, Bangladesh, Cambodia, India, Lao People's Democratic Republic, Nepal and Pakistan. In South Asia, 40 per cent of the population lives on less than US$1 a day.

Poverty is not just determined by economics, however. Conventional indicators suggest that many Pacific Island populations are at poverty level - however, many communities still enjoy a high degree of subsistence affluence from traditional, non-monetary resource management systems. Poverty for many countries of the region is the result of gross inequalities and institutional failures which allow the benefits of economic growth to be captured by a decreasing number of elites. In addition, urbanization, the shift to a monetary economy and high population growth without commensurate growth in employment opportunities compound the problem.

Nevertheless, the region includes some of the most populous countries in the world, with China and India accounting for 38 per cent of the world population. It also contains five of the six countries which account for one-half of global annual population growth - Bangladesh, China, India, Indonesia and Pakistan. High population growth is reflected in the region's population structures. Most countries have youthful populations, with 30 per cent of Asia's population less than 15 years old.

While this could be seen as a positive characteristic in terms of the large number of young workers available, in some sub-regions, especially the Pacific Islands, it also has significant negative socioeconomic implications, particularly in terms of high unemployment. In addition, large numbers of young people entering their reproductive years compound the pressures of population growth. Despite gains in life expectancy, an estimated 7.1 million people live with HIV/AIDS in Asia and the Pacific. There were about 435 000 deaths and more than 1 million new cases in 2001.

During the past 30 years, countries in the region have striven for economic development and higher standards of living. However, annual GDP growth rates decreased from a high of 9.76 per cent in 1970 to 2.54 per cent in 1999, with a negative growth of -1.04 per cent in 1998 due to the Asian economic crisis. In South Asia, the growth rate also exceeded 2 per cent. However, growth was very low in the Pacific Islands, and this is consistent with recent studies that indicate a general decline in the standard of living in Pacific island countries.

The economic structure of the region has changed remarkably over the past 30 years, with the importance of agriculture diminishing and the service sector growing. Even in South Asia, the contribution of agriculture to GDP fell from 39 per cent in 1980 to 30 per cent in 1995 while the contribution of the service sector increased from 35 to 41 per cent. These structural changes are also reflected in employment. In 1960, 75 per cent of Asians were employed in agriculture.

In the Pacific, lifestyles have changed from subsistence to cash-driven societies reliant on budgetary assistance. The standard of living for the sub-region's urban dwellers is relatively high when compared with those in other developing countries. However, there are some worrying trends, with indications of rising unemployment, particularly among young people; high drop-out rates from primary schools; low household cash incomes; and a growing incidence of drug abuse and crime. Many of the small and remote islands of the Pacific have essentially no industry at all, while other countries in the sub-region have small industries related to food or beverage processing, clothing and minor machinery assembly or repair.

The region has at least ten global hubs of technological innovation in Australia, China, India, Japan, Malaysia, Republic of Korea, Singapore and Taiwan. Asia accounts for about 30 per cent of global expenditure on research and development, with Japan alone accounting for half of that. In line with developments in other parts of the world, the spread of new technologies such as the Internet and mobile communications has been unprecedented and is having significant impacts on people's lives and selected national economies. The glowing picture of Asia's future in the early to mid-1990s has been overshadowed by more recent events in Southeast Asia and Korea.

There has been a loss of confidence in the region which has ensured that the region's leaders are now more attentive to the need for adaptive governance and fiscal management to guard against future downturns. For countries to prosper again, governments and institutions must encourage new or burgeoning markets and pursue social policies that will simultaneously benefit the economy, the environment and the people.

The most important developments in Europe over the past three decades are the political, economic, social and institutional processes resulting from the strengthening and expansion of the European Union (EU) and the transition from centrally planned regimes to more open, market economy based societies. These changes have had profound effects on developments in all the countries concerned, on sub-regions and on the region as a whole. Although the three sub-regions of Europe do have similarities, there also are distinct differences due to recent and historical events resulting in political, economic and social heterogeneity in the region.

Europe is predominantly a region of high to medium levels of human development. However, while the overall level continues to improve gradually throughout Western and parts of Central Europe, many countries of Eastern Europe have suffered severe setbacks, including a rise in income poverty, since the beginning of the transition process. The region traditionally has high rates of adult literacy, estimated at 95 per cent or more for Europe as a whole, although rates tend to be slightly lower in southern parts of Western Europe. In several of the CEE countries half or more of the population had incomes below the official poverty line in the period 1989-95.

This impoverishment is reflected in a drastic fall in real wages and per capita GDP, high rates of inflation and a rise in income inequalities - including between men and women, the latter often being the first to lose their jobs. Relative prices have also changed, with prices of goods and services needed by the poor often rising much faster than others. While income poverty is clearly more pervasive and severe in Eastern Europe, it is not unknown in Western Europe, with an estimated 17 per cent of the EU population still experiencing poverty.

Vulnerability to income poverty is more widespread: 32 per cent of Europeans experience at least one annual spell of low income over a period of three years, while 7 per cent experience persistent poverty during this period. The human costs of the transition process have reached beyond income poverty alone. In Europe as a whole, life expectancy has increased in the period 1995-2000 compared to 1975-1980 from 70.3 to 73.1 years. However, in some Eastern European countries life expectancy has decreased over the same period, especially for men - for example from 62 to 58 in the Russian Federation and from 65 to 64 in Ukraine.

In addition, in many CEE countries, the ratio of men to women is far below the standard ratio. The causes of this 'missing men issue' are multiple and complex but stem mainly from human insecurity: military conflict, poor health, unemployment, loss of pensions and corruption, all of which result in social breakdown and a poor quality of life. Dismantling of the communist era welfare system also led to social disintegration and inequality in social services in CEE. This decline was associated with a proliferation of fraud, illegal businesses and organized crime. In sharp contrast to conditions before transition, people now find themselves deprived of personal safety and security - often at the mercy of organized criminal forces that have arisen on the basis of collusion with corrupt government officials. The increase in crime reveals a weakness in state authority and in public law enforcement.

Latin America and the Caribbean has experienced financial and political turmoil over the past three decades. While some aspects of human development have improved significantly, poverty - especially in rural populations - and inequity persist as major problems, hindering regional efforts to move towards sustainable development. Poverty is widespread. It is estimated that about 200 million people, or 40 per cent of the region's population, live in poverty.

The incidence of poverty is higher in rural areas but there are more poor people in urban areas than in rural ones, and almost one-half of the poor are children or youth. Life expectancy increased from 65.8 to 72.5 years between 1970 and 2000, with important national and sub-national differences that are related to per capita income levels. The highest life expectancies are in the Caribbean and in South America, although there are variations with sub-regions - life expectancy is 20 years longer in Cuba and Puerto Rico than in Haiti and 10 years longer in Venezuela and Colombia than in Bolivia.

Despite this, all countries except Haiti now exceed the life expectancy goal of 60 years proposed for the region in 1977 as part of the World Strategy on Health for All by the Year 2000. More efficient health programmes have also lowered the child mortality rate significantly, from 81.6 per 1,000 births in 1970 to 35.5 per 1,000 in 1995. Education has also improved in the past two decades. Adult literacy rates are generally high at about 88 per cent in 1999, a leap from 77 per cent in 1980.

However, the highly unequal distribution of income, which is pervasive throughout the region, is mirrored by inequalities in access to schooling, attendance and performance. Violence, including homicide, is on the increase in the entire region, particularly in Colombia and Brazil. Furthermore, violence within the family has increased, particularly towards women and children; it is estimated that about half the women in Latin America face at least one episode of family violence in their life.

The past three decades of the 20th century have brought increasing affluence and power to North America. North Americans not only live long lives in increasingly diverse societies but their production of material wealth and consumption of goods are also among the highest in the world. American capital, technology and goods are fuelling globalization, a defining trend of the new millennium that carries unprecedented opportunities and risks.

Based on the Human Development Index (HDI), the region probably has the highest level of human development in the world. With Canada ranked third and the United States sixth, the region has an average HDI value of 0.935, compared to 0.928 for the highincome OECD countries. Despite this, poverty is not unknown in North America. Over the past decade, poverty rates have declined in the United States but increased in Canada. Although debates about the definition, measurement and thus extent of poverty continue, data clearly show that some social groups are more vulnerable than others. Poverty is more likely to affect aboriginal people, some minorities, single parents and children.

Environmental concern was increased during the mid-1980s by a new awareness of the global nature of some environmental problems, and membership in environmental NGOs soared. By the 1990s, 'common sense' approaches were advocated as concerns over deficit reduction led to budget cuts to environmental departments and to reliance on market incentives and voluntary programmes. After the UN Conference on Environment and Development in 1992, both countries became committed to sustainable development as reflected in stated federal goals in Canada and the efforts by many US states and localities in moving forward on Agenda 21.

The West Asia region has undergone major demographic development and socio-economic transformation, including substantial agricultural and industrial development, since oil was discovered at the beginning of the 20th century. These trends have intensified during the past 30 years. Most West Asian countries fall into the high (Bahrain, Kuwait, Qatar and the United Arab Emirates) or medium (Jordan, Lebanon, Oman, Saudi Arabia and Syria) category of human development. Yemen is the only country classified as having a low level of human development, and information is not available for Iraq and the Occupied Palestinian Territories.

Human development rankings were higher for most countries in the 1990s than the early 1980s although many have experienced set-backs over this period, particularly in the 1990s. In some countries, there has been a significant improvement in certain components of human development over the past three decades. For instance, in Oman life expectancy increased from 54.9 years in the early 1970s to more than 70 in 2000 although in Iraq it declined from 66 to 58 years over the same period.

Access to improved water and adequate sanitation is generally high with the exception of Yemen, where the figures are 69 per cent for improved water and 45 per cent for adequate sanitation. Access to safe water also dropped in Iraq during the second half of the 1990s. There is considerable variation in per capita GDP between the countries of West Asia. The highest are on the Arabian Peninsula, ranging from US$6 384 in Saudi Arabia to US$16 483 in Kuwait in 1998.

Seven West Asian countries (Iraq, Jordan, Lebanon, Oman, Saudi Arabia, Syria and Yemen) score poorly on one or more of the key components used to assess levels of human development. For most of these countries, adult literacy and low life expectancy are the key factors, rather than income poverty which is mainly an issue in Jordan, Oman and Yemen.

3

Environment and Human Vulnerability

Human vulnerability integrates many environmental concerns. Since everyone is vulnerable to environmental threats, in some way, the issue cuts across rich and poor, urban and rural, North and South, and may undermine the entire sustainable development process in developing countries. Reducing vulnerability requires identifying points of intervention in the causal chain between the emergence of a hazard and the human consequences. Many natural phenomena pose threats, including extreme events such as floods, drought, fire, storms, tsunami, landslides, volcanic eruptions, earthquakes and insect swarms.

Human activities have added to the list, with threats from explosions, chemical and radioactive contamination, and other technological incidents. The risk lies in the probability of exposure to any of these events, which can occur with varying severity at different geographical scales, suddenly and unexpectedly or gradually and predictably, and to the degree of exposure. With an increasing and more widely distributed global population, however, natural disasters are resulting in increasing damage, loss of life and displacement of populations.

In addition, human-induced changes to the environment have reduced its capacity to absorb the impacts of change and to deliver the goods and services to satisfy human needs. Environmental change and social vulnerability to it is nothing new. More than 9,000 years ago, the Sumerians of Mesopotamia started irrigating land to meet increased demand for food from a growing population but their civilization eventually collapsed partly because of the waterlogging and salinization that resulted. The Mayan civilization collapsed around 900 B.C. mainly as a result of soil erosion, loss of agro-ecosystem viability and silting of rivers. The Dust Bowl phenomenon of the American prairies in the 20th century resulted from massive soil erosion, and led to communities being uprooted and widespread poverty.

Some people live in places of inherent risk to humans - areas, for example, that are too hot, too dry or too prone to natural hazards. Others such as Rosita Pedro are at risk because an existing threat has become more severe or extensive through time. Places or conditions which were once safe have been so altered that they no longer safeguard human health and well-being adequately. Many of the children under the age of five who die every year from diarrhoeal disease contract it from drinking contaminated water.

Most environments are in a constant state of flux because of natural causes and human modifications for food production, settlements, infrastructure, or to produce and trade goods. Most intentional changes are designed to harness the environment for human benefit. Domestication of land for intensive food production is one example; harnessing river resources to provide fresh water, energy and transport is another.

Such changes may also unintentionally alter the quality or quantity of environmental resources and be difficult to cope with. Analysing old and new threats to human security shows that human vulnerability to environmental conditions has social, economic and ecological dimensions. The most conspicuous and widely reported manifestation of this vulnerability is when people are affected suddenly and violently by natural hazards such as the eruption of Mount Nyiragongo resulting in the devastation of the town of Goma in the Democratic Republic of Congo. These events turn into disasters when local communities are not able to cope with their impacts.

The environmental factors that contribute to human vulnerability, however, are both varied and variable, and are not limited to disaster events; they span the whole sustainable development spectrum. Although everyone is vulnerable to environmental impacts of some kind, the ability of people and societies to adapt to and cope with change is very varied. Developing countries, particularly the least developed, have less capacity to adapt to change and are more vulnerable to environmental threats and global change, just as they are more vulnerable to other stresses. This condition is most extreme among the poorest people and disadvantaged groups such as women and children.

The coping capacity of human society is a combination of all the natural and social characteristics and resources available in a particular location that are used to reduce the impacts of hazards. These include factors such as wealth, technology, education, information, skills, infrastructure, access to resources and management capabilities. Between two and three times as many disaster events were reported in the United States in 1999 as in India or Bangladesh but there were 14 times and 34 times more deaths in India and in Bangladesh, respectively, than in the United States. The critical factor behind these statistics lies in the advantages enjoyed by US citizens in terms of levels of coping capacity. There is, therefore, no direct correlation between the occurrence of extreme events and their level of human impact.

In many instances, coping capacity that was adequate in the past has not kept pace with environmental change. This can happen when traditional options are reduced or eliminated, or when new threats emerge for which no coping mechanism exists, resources are lacking, and technology and skills are not available. Some groups are more exposed than others to particular environmental risks: urban populations are exposed to high levels of contaminant and particulate pollution in the air, slum dwellers often lack the minimum protective infrastructure, employees may be exposed to particular hazards in the work place, and the uninformed may simply not know about the threats that surround them.

A wide range of social and economic factors have direct and indirect bearing on human vulnerability to environmental change, including poverty and inequality, and the availability of natural resources. No standard framework exists for identifying all these factors. Poverty is generally recognized as one of the most important causes of vulnerability to environmental threats, on the basis that the poor tend to have much lower coping capacities, and thus they bear a disproportionate burden of the impact of disasters, conflict, drought, desertification and pollution.

But poverty is not the only reason. The very young and the old, women and children are often identified as especially vulnerable groups. Refugees, migrants and other displaced groups lack both the physical resources and social structure necessary to respond to threats although paradoxically they may initially benefit from the high visibility of their plight. The urban poor, on the other hand, usually live in obscurity, and in times of disaster their numbers can swell enormously.

The mosaics of vulnerability seem so complex as to cast doubt on attempts to describe patterns and estimate trends at the global or even the regional scale. General or gradual economic decline can affect vulnerable groups disproportionately, creating severe but largely hidden hardships. The cultural dimension is important. Indigenous communities with unique lifestyles intimately adapted to local climate, vegetation and wildlife may be particularly threatened by environmental change.

Traditionally, many indigenous communities developed highly specific coping mechanisms to deal with their environments and periodic extreme events. Such coping mechanisms included adaptive behaviour such as regular seasonal migration or exceptional relocation in times of flooding or drought, and changes in practices such as planting and gathering specific food crops; for example, fruits and foods that are not usually eaten during times of good harvests may be relied upon in times of crop failure.

With the breakdown of social patterns, and reduction of options to continue following indigenous lifestyles, such coping mechanisms are also giving way or disappearing. Poor and indigenous communities are considered to be more vulnerable to climate-related events such as storms, floods and droughts because of inadequacies in social support services and systems such as water management infrastructure. They are also more affected by pests and diseases - especially vector-borne, respiratory and other infectious diseases.

In addition, since many poor inhabit isolated rural environments or the margins of large towns and cities, they are more exposed to social problems associated with economic insecurity, inadequate water supplies and lower health standards. Human exposure to environmental threats is not evenly distributed. Some locations, such as high latitudes, floodplains, river banks, small islands and coastal areas, may pose more risk than others. Human uses or modifications of the environment such as deforestation, increasing paved areas covered by buildings and roads, and river canalization have created impacts that often affect areas a long way from the source of the environmental change, such as far downstream.

Individual choices have an enormous bearing on where people live and work, with the result that human vulnerability is closely related to population density and distribution. Floodplains, low-lying coastal areas and volcanic areas have always been favoured for settlement because of their soil fertility or the availability of flat land. As populations increase and there is more competition for land and resources, areas of higher potential risk are increasingly being settled, such as mountains, steep slopes and locations near sources of pollution.

Such settlers are vulnerable to the associated single or combined hazards such as landslides, flooding, volcanic eruptions and toxic chemicals. Again, the poorest strata of society are often the most vulnerable because they have fewer options in where to live. For various reasons, even the more affluent often choose to live or work in areas prone to environmental threats or hazards. Those living along the earthquakeprone San Andreas fault in California are a prime example, as are those who settle in hurricane belts, on sand spits, on eroding coastlines or in towns where water supplies are inadequate to meet demand.

Clearly, the benefits of the location (employment, job security, leisure facilities) are perceived to outweigh the known risks. Measures to mitigate the risks may be sought in the form of insurance or purchasing a scarce commodity such as water but these options are not always appropriate, available or affordable to all members of the community. In 2002, more than 1 billion urban dwellers, mostly in Africa, Asia and Latin America, live in slums or as squatters. Of the projected 1 billion new urban dwellers by 2010, most will probably be absorbed by cities in developing countries that already face multiple problems such as shortages of adequate housing, infrastructure, potable water supplies, adequate sanitation and transportation systems as well as environmental pollution.

The urban poor, unable to afford alternatives, are frequently forced to live in areas with the worst urban services and most unhealthy environmental conditions, exposed to multiple hazards and increased risk, their vulnerability enhanced by overcrowding. Some communities have become more vulnerable because the scarcity of critical resources such as land, fresh water and forests is contributing to conflicts. These environmental scarcities do not usually cause wars among countries but they can generate severe social stresses within countries or across borders, helping to stimulate sub-national insurgencies, ethnic clashes and urban unrest. Such civil violence affects developing societies particularly because they are generally more dependent on environmental resources and less able to buffer themselves from the social crisis that environmental scarcities cause.

Environmental Change

Two basic functions performed by the environment are the 'source' or production function that supports the livelihood of millions who depend upon environmental resources, and the 'sink' or pollution absorption and cleansing function essential for human health and well-being. Not only are these two functions closely connected in a cycle of production and renewal but they are being increasingly impaired and degraded by human impacts. Degradation of natural resources such as land, fresh and marine waters, forests and biodiversity threatens the livelihood of many people but especially the poor.

For example, water tables are falling fast under the North China plain. In 1997, almost 100 000 wells were abandoned apparently because they ran dry as the water table fell, but 221 900 new wells were drilled. The drilling of so many wells reflects a desperate quest for water. The 'sink' function of the environment operates through such processes as nutrient recycling, decomposition, and the natural purification and filtering of air and water. When these functions are impaired, health can be jeopardized by contaminated household water, sanitation problems, indoor air pollution, urban air pollution and agrochemical pollution.

Environmental change may have impacts on health, habitat and infrastructure, economy, society and culture, increasing vulnerability. Human health is increasingly determined by environmental conditions. According to a report from the World Health Organization, for example: Deteriorating environmental conditions are a major contributory factor to poor health and poor quality of life. Mismanagement of natural resources, excessive waste production and associated environmental conditions that affect health pose major challenges to sustainable development.

Impoverished populations living in rural and periurban areas are at greatest risk from degraded environmental conditions. The cumulative effects of inadequate and hazardous shelter, overcrowding, lack of water supply and sanitation, unsafe food, air and water pollution, and high accident rates, have serious effects on the health of these vulnerable groups. Poor environmental quality is directly responsible for some 25 per cent of all preventable ill health, with diarrhoeal diseases and acute respiratory infections heading the list.

Two-thirds of all preventable ill health due to environmental conditions occurs among children. Air pollution is a major contributor to a number of diseases, and to a lowering of the quality of life in general. There are regional differences in the way human health is vulnerable to environmental degradation. Communities in many parts of Central and South America, Central Africa and Asia are highly vulnerable to water-borne and vector-borne diseases.

Air pollution threatens large urban areas and mega-cities, most of which are in developing countries. People in developed countries are more vulnerable to exposure to toxic chemicals and technological accidents but there are notable exceptions such as arsenic contamination in south Asia. Overall, it is estimated that 25-33 per cent of the global burden of disease is attributable to environmental factors. Recent estimates suggest that environment-related premature death and illness account for 18 per cent of the total burden of disease in the developing world.

This comprises contributions from water supply and sanitation, indoor air pollution (4 per cent), vector-borne diseases (3 per cent), urban air pollution (2 per cent) and agro-industrial waste (1 per cent). In sub-Saharan Africa the figure is even higher at 26.5 per cent, mainly related to water supply and sanitation (10 per cent) and vector-borne diseases (9 per cent). Globally, 7 per cent of all deaths and diseases are due to inadequate water, sanitation and hygiene. Approximately 5 per cent are attributable to air pollution.

Every year, environmental hazards kill 3 million children under the age of five. Current estimates suggest that 40-60 per cent of those deaths are due to acute respiratory infection resulting from environmental factors, particularly particle emissions from solid fuel use. In the United States, a 10 µg/m3 increase in fine particle air pollution results in a 4 per cent increase in general morbidity, a 6 per cent increase in cardio-pulmonary mortality and an 8 per cent increase in lung cancer mortality.

In the short term, disease due to environmental change is likely to have more impact on developing countries than on developed ones. This is partly because developed countries have devoted considerable effort to reducing the health threat from dirty water, poor sanitation and using solid fuels in open fires inside homes. The same is not true for most developing countries. As a result, exposures to particulates for non-smokers are often an order of magnitude lower in developed countries than in developing ones.

In Helsinki, for example, the particulates in the air come mainly from indoor dust, cleaning products, traffic and long-range transport. In developing countries, the use of solid fuel as a primary energy supply dominates the exposure of non-smokers to particle pollution, especially among women and children in rural and slum environments. In the past decade, smoke haze from forest fires has also become an important source of respiratory disease. Furthermore, most developing countries still lack the resources to deal effectively with public health crises and are situated in regions where many water-borne and vector-borne diseases are acute.

Microbiological contamination of the sea by sewage pollution has precipitated a health crisis of massive proportions globally. Bathing in polluted seas is estimated to cause some 250 million cases of gastroenteritis and upper respiratory disease every year, with an estimated annual cost worldwide of about US$1.6 billion. Some of these people will be disabled over the longer-term, suggesting that the global impacts of marine pollution are comparable to those of diphtheria and leprosy. Eating sewage-contaminated shellfish causes an estimated 2.5 million cases of infectious hepatitis a year, of whom some 25,000 die and another 25,000 suffer longterm disability resulting from liver damage. The annual global burden on human health is estimated to equal some 3.2 million DALYs - comparable to the worldwide impact of all upper respiratory infections and intestinal worm diseases - and to cost world society some US$10 billion annually.

Food Security

Food security means being able to obtain a nutritionally adequate, culturally acceptable diet at all times through local non-emergency sources. This requires both adequate food production or imports, and economic access to food at the household level, at all times, to ensure a healthy active life. This idea goes well beyond the traditional concept of hunger: it embraces a systematic view of the causes of hunger and poor nutrition within a community, recognizing both physical and economic vulnerability.

Projections of production increases suggest that the global availability of food should be adequate in coming decades. Aggregate statistics, however, are often misleading, and can hide the real situation on the ground. For example, per capita food production in Africa has declined slightly over the past 30 years and decreased significantly in the former Soviet Union since 1990. Agricultural growth as a consequence of the Green Revolution has also had an adverse impact on the environment in terms of nutrient mining, increase in soil salinity, waterlogging, depletion of underground water and the release of nitrogen into watercourses.

Human vulnerability to environmental change has an important economic dimension. Human well-being is inextricably linked to ecosystems through the goods and services that ecosystems provide. This includes both marketed goods and services, such as food or forest products, and non-marketed ones such as water flow regulation, so that any reduction or degradation in supply leads to a loss of human welfare. In Japan, for example, the damage to agricultural crops caused by tropospheric ozone amounts to an estimated US$166.5 million yearly in the Kanto region alone.

The economic dimensions of vulnerability to environmental change often focus on the impact of natural disasters or other extreme events. While total losses may be highest in developed countries, with their expensive infrastructure, the impact on the economies of developing regions may be greater. For example, the 1991-92 drought that hit most of Southern Africa resulted in a decline of 62 per cent in the Zimbabwe Stock Market.

The potential economic losses of non-marketed ecosystem goods and services and the impact on human vulnerability are likely to be even higher than for marketed goods and services. Equally, little attention is paid to the high economic cost of more gradual environmental degradation and loss of natural resource potential. The cumulative evidence for increasing human vulnerability to environmental change calls for a significant policy response and action on several fronts.

Social responses have frequently focused on 'downstream' measures, designed to mitigate the hardships and cushion the impacts of environmental change or natural disaster after the event, rather than on interventions intended to modify basic driving forces ahead of a potential crisis. The onset of conditions that give rise to threats and vulnerability can often be gradual or inconspicuous. Donors are often ready to offer relief once a high-profile disaster such as a famine or flood has occurred but they are less likely to finance precautionary measures. 'Upstream' intervention is generally highly costeffective and should be given greater priority.

Levels and trends of vulnerability for different groups need to be assessed regularly as a basis for designing specific measures for vulnerability reduction and evaluating their impact. Governments need to assess and map national threats due to environmental change, particularly those that may be growing, and to institute early warning, mitigation and response measures to reduce the human and economic costs of disasters that are in part avoidable. Vulnerability should be recognized as a key indicator of the seriousness of environmental problems such as global warming. It should be a focus for developing policies that seek to help people avoid, cope with or adapt to adverse effects of environmental change.

Climate Change and Poverty

Climate change will compound existing poverty. Its adverse impacts will be most striking in the developing nations because of their geographical and climatic conditions, their high dependence on natural resources, and their limited capacity to adapt to a changing climate. Within these countries, the poorest, who have the least resources and the least capacity to adapt, are the most vulnerable. Projected changes in the incidence, frequency, intensity, and duration of climate extremes (for example, heat waves, heavy precipitation, and drought), as well as more gradual changes in the average climate, will notably threaten their livelihoods - further increasing inequities between the developing and developed worlds.

Climate change is therefore a serious threat to poverty eradication. However, current development strategies tend to overlook climate change risks. An approach that uses both mitigation and adaptation is needed. Current commitments to mitigate climate change by limiting the emissions of greenhouse gases (GHGs) will not, even if implemented, stabilize the atmospheric concentrations of these gases.

Developing adaptive capacity to minimize the damage to livelihoods from climate change is a necessary strategy to complement climate change mitigation efforts.

Climate change adaptation - all those responses to climatic conditions that reduce vulnerability - is therefore an integral and urgent part of overall poverty reduction strategies. Adaptation should not be approached as a separate activity, isolated from other environmental and socioeconomic concerns that also impact on the development opportunities of the poor.

A comprehensive approach is needed that takes into account potential synergistic and antagonistic effects between local and global environmental changes as well as socioeconomic factors. Today, it is widely agreed by the scientific community that climate change is already a reality. The rate and duration of warming observed during the twentieth century are unprecedented in the past thousand years. Increases in maximum temperatures, numbers of hot days, and the heat index have been observed over nearly all lands during the second half of the twentieth century.

There is increasing observational evidence that regional changes in climate have contributed to various changes in physical and biological systems in many parts of the world. These include the shrinkage of glaciers, thawing of permafrost, changes in rainfall frequency and intensity, shifts in the growing season, early flowering of trees and emergence of insects, and shifts in the distribution ranges of plants and animals in response to changes in climatic conditions. On the regional level, climate change is superimposed on the existing climatic conditions and manifests itself through:

— Changes in average climatic conditions. For example, some regions may become drier or wetter on average.

— Changes in climate variability. For example, rainfall events may become more erratic in some regions.

— Changes in the frequency and magnitude of extreme events.
— Changes in sea levels, which are projected to rise by between 0.09 and 0.88 meters by 2100 relative to 1990.

The impacts of climate change vary across geographical regions. Some of the anticipated impacts of climate change are positive. For example, waterscarce regions such as parts of Southeast Asia may benefit from increased water availability. However, developing countries are likely to suffer most from the negative impacts of climate change. This is due to the economic importance of climate-sensitive sectors for these countries, and to their limited human, institutional, and financial capacity to anticipate and respond to the direct and indirect effects of climate change.

In general, the vulnerability is highest for least developed countries (LDCs) in the tropical and subtropical areas. Hence, the countries with the fewest resources are likely to bear the greatest burden of climate change in terms of loss of life and relative effect on investment and the economy. The extent and scope of regional climate change impacts depend on the degree of mitigation. While the urgency and scale of adaptation efforts required will be lower if aggressive mitigation is undertaken early on, some degree of adaptation is inevitable. Reductions in emissions of greenhouse gases would delay and reduce damages caused by climate change.

Essentially, the lower the future stabilization level of atmospheric greenhouse gas concentration, the less would be the likely damage. The UNFCCC states that: "the parties should protect the climate system for the benefit of present and future generations of humankind, on the basis of equity and in accordance with their common but differentiated\ responsibilities and respective capabilities.

Accordingly, the developed\ country Parties should take the lead in combating climate change and the adverse effects thereof." (Article 3.1 of the UNFCCC). Even if greenhouse gas emissions were curbed immediately, the global average temperature would still continue to rise due to the slow response of the Earth's atmosphere system to past emissions. This suggests that any future levels of greenhouse gas concentration, once stabilized, will be above current levels. Before addressing climate change adaptation in the context of development, it is important to recognize that today's climate already influences economic opportunities and development prospects.

Poor countries and people tend to be particularly vulnerable to deviations from average climatic conditions and climatic extremes. Climate and climate variability are therefore important elements of the complex web of factors influencing people's livelihoods. When comparing data on natural hazards in developing and developed countries, the loss of life and the number of people affected tend to be considerably larger in developing country regions for natural disasters of comparable magnitude. Damages in relation to GDP are usually also higher.

Bangladesh is a prime example of a country that is particularly vulnerable to today's climate. With a low-lying coastline, high population density, and an economy highly dependent on agriculture, the lives and livelihoods of people are threatened by frequent cyclones and the associated effects, such as saltwater intrusion, that render agricultural lands unproductive. Between 1974 and 1998, the country experienced seven major floods. In 1998, about 68 percent of the country's geographical area was flooded, affecting more than 30 million people and causing 918 fatalities.

Economic losses were estimated at US$3.3 billion, equivalent to 8 percent of the country's GDP. The impact of climate variability on countries is also well illustrated by the environmental and socioeconomic damages associated with El Niño. El Niño is a natural recurring climatic phenomenon associated with fluctuations in the atmospheric pressure and sea surface temperatures in the tropical Pacific Ocean. It affects the climate on a global scale, with the impacts concentrated in the tropical and subtropical regions. The shift in sea surface temperature is known to affect marine productivity.

On land, El Niño is associated with floods and droughts in Latin America, Asia, and Africa, as well as changes in extreme events and the distribution of vector-borne diseases. El Niño has caused loss of life, destroyed livelihoods, and affected national economies. For Ecuador, the overall costs of direct damages to agriculture, livestock and fishing associated with the 1997-98 El Niño constituted 4.7 % of its agricultural GDP. Loss of harvest and rising unemployment of agricultural workers led to an increase in the incidence of poverty by 10 percentage points in the affected municipalities.

Between 1970 and 1999 about 3.76 billion people were affected by natural disasters in Asia, explained in a large part by Asia's high population density in hazard prone areas. Africa had the second highest number of people affected by natural disasters, largely due to frequent occurrence and the longterm effects of droughts and the importance of the agricultural sector. For the regions of Latin America and the Caribbean, floods had the highest cumulative cost, followed by windstorms, earthquakes, and droughts.

All societies and economies have developed mechanisms to cope with climate extremes and other natural hazards, which they face occasionally. Trade, migration, or precautionary storage of food are examples of strategies to cope with adverse climatic conditions. This capacity to cope with climate variability and extreme weather events in itself is highly dependent on the level of economic development. In general, livelihood sources of the poor are usually narrower and more climate-sensitive than those of the non-poor.

Extreme weather events, which would cause limited damage and few casualties in a developed country, often cause extensive damage and substantial loss of life in a developing country. Poor people are particularly vulnerable to deviations from average climatic conditions such as prolonged drought and to natural disasters such as floods. In periods of stress they may be forced to sell off their physical assets such as land, bicycles, and farming implements, thereby undermining the sustainability of their livelihoods over the longer term.

Among the poor, vulnerability varies, since some groups are more lacking in the financial, social, and political means of securing alternative livelihoods less exposed to risk than others. Women for example may be constrained by social and cultural structures that place them in inferior social positions, limiting their access to income, education, public voice, and survival mechanisms. In addition, the coping capacities of the poor are often already strained due to a number of trends including HIV/AIDS, increasing population densities, and detrimental forces associated with globalization. Climate change will add to these trends and increase vulnerabilities.

Risks and Vulnerabilities

In the face of changing patterns of climate variability, and significant deviations from historical experience, their effectiveness may be significantly reduced. For example, in Tanzania, high rainfall due to the 1998 El Niño was followed by a two-year period of erratic rainfall. This climatic shock caused some of the poorer farmers to give up maize farming and opt instead to sell their labor at farms in other, more productive areas. The resulting dependence on physical working capacity as their sole endowment increased vulnerability, since malnutrition and disease can reduce their capacities for manual labor.

Since the mid-1970s El Niño events have become more frequent, persistent and intense than the opposite cool phase. Whether this already the result of climate change is the subject of ongoing scientific debate. However, such deviations from normal climatic conditions and previous experience illustrate the additional strain climate change is likely to exert on the poor, if no appropriate adaptation measures are taken.

The poor will need to devote more of their already limited resources to coping with adverse climatic conditions. Climate change may thus force drastic changes to livelihood strategies. Where economic diversification is low, income opportunities and hence options for developing alternative livelihoods in response to climatic changes may be limited. In some cases migration, which is an important coping strategy for poor people, might be the only solution, but will potentially cause social disruption.

The impacts of climate change on the poor will be context-specific, reflecting factors such as geographic location; economic, social, and cultural characteristics; prioritization and concerns of individuals, households, and social groups; as well as institutional and political constraints. The following points illustrate the impacts of climate change on poor people's livelihoods.

Ecosystem Goods and Services

The degree of local environmental degradation will influence the vulnerability of an ecosystem to climate change. Habitat fragmentation is already a leading cause of biodiversity loss and changes in temperature and moisture regimes further limit habitats necessary for the survival of species. Degradation of forested mountain slopes in conjunction with intensified rainfall may increase erosion and loss of fertile soil and affect the quality of watersheds.

Climate change is likely to lead to changes in species distribution and abundance, and increase the risk of extinction and loss of biodiversity. Since some ecosystems are highly sensitive, even small changes can have large effects. Minor increases in water temperature can, for example, damage coral reefs, exacerbating other stresses such as pollution and over-fishing and thereby cause a reduction in fish stocks, jeopardizing fish- and tourismdependent livelihoods. Poor people are often directly dependent on goods and services from ecosystems, either as a primary or supplementary source of food, fodder, building materials, and fuel. This makes them highly vulnerable to ecosystem degradation.

While local economic and social conditions drive poor people into marginal areas and force them to exploit natural resources to support their livelihoods, climate change further erodes the quality of the natural resource base, thereby reinforcing conditions of poverty. Changes in ecosystem composition and provision of goods and services may also have wider economic effects. Essential ecosystem services include breaking down wastes and pollutants, purifying water, and maintaining soil fertility. Climate change will alter the quality and functioning of ecosystems, reducing their capacity to perform their role as important life support systems. This will have important impacts on key economic sectors such as agriculture, water supply, and others.

Water

Water scarcity is already a major problem for the world's poor. The number of people impacted by water scarcity is projected to increase from about 1.7 billion people today to around 5 billion people by 2025, independent of climate change. Climate change is projected to further reduce water availability in many water scarce regions, particularly in the subtropics, due to increased frequency of droughts, increased evaporation, and changes in rainfall patterns and run-off. Precipitation is expected to increase in equatorial, middle, and high latitude regions, which tend to suffer less from water scarcity.

As rainfall events are expected to become more intense, the incidence of floods may increase, jeopardizing human settlements and infrastructure. Increases in temperature and changes in precipitation are projected to accelerate the retreat and loss of glaciers. Associated changes in the timing of streamflow will have downstream effects for agriculture. The melting of glaciers has become a serious concern in the Himalayan region, because of the growing risk of glacial lake out-burst floods.

Agriculture and Food Security

Agriculture is the most important sector for most least developed countries as the impact of agricultural growth on poverty reduction tends to exceed the impact of growth in other sectors. Food security is a function of several interacting factors, including food production as well as food purchasing power. Climate change could worsen the prevalence of hunger through direct negative effects on production and indirect impacts on purchasing powers.

Land degradation, price shocks, and population growth are already a major concern for sustaining agricultural productivity. Changes in temperature, precipitation, and climatic extremes will add to the stress on agricultural resources in many developing country regions and reduce the quality of land areas for agricultural production. This will be particularly serious for areas where droughts and land degradation, including desertification, are already severe. As access to productive land is important for reducing rural poverty, the impacts of climate change on the productivity of land will further constrain efforts to combat rural poverty.

Low-lying coastal communities will have to deal with sea level rise and the impact of climate change on marine resources. Sea level rise may lead to salinization and render agriculture areas unproductive. In areas where fish constitute a significant source of protein for poor people, declining and migration of fish stocks due to climate change and associated changes in the marine environment will further need to be considered in their impact on the local food security.

The impact of climate change on food supply varies significantly by region. In general, crop yields are projected to decrease in most tropical and subtropical regions due to changes in temperature and rainfall. Consequently, there is a real risk that climate change will worsen food security and exacerbate hunger in some developing-country regions. In the short term, however, the greater impact on food security could come from the projected increases and severity of extreme weather events rather than from gradual changes in the climate.

The impact of climate change on food security will be a major concern for Africa. In conjunction with the previously discussed changes in water supply, the production losses for Sub-Saharan countries could be substantial as the length of suitable growing periods decreases. Livestock activities and crop yields for many countries in Asia and Latin America are also projected to decrease.

Health

The potential impacts of climate change on human health would increase vulnerability and reduce opportunities by interfering with education and the ability to work. While any attempt at predicting and gauging the impact of climate change on human health is a complicated task, it is likely that climate change will have both direct and indirect adverse effects on human health.

A direct effect is an increase in temperature-related illnesses and deaths. Prolonged intense heat waves coupled with humidity may increase mortality and morbidity rates, particularly among the urban poor and the elderly. Another direct effect will be increased death and injury from extreme weather events such as flooding, landslides, and storms -over 96 percent of disaster-related deaths in recent years have taken place in developing countries. Changes in temperature and rainfall may change the geographic range of vector-borne diseases such as malaria and dengue fever, exposing new populations to these diseases.

Young children as well as pregnant women and their unborn children are especially vulnerable to malaria. Malaria contributes to perinatal mortality, low birth weight, and maternal anemia. The frequency and severity of malaria epidemics in East Africa already appear to have increased in correspondence with the increased frequency, magnitude, and persistence of the El Niño phenomenon over the past 20 to 30 years. The net effect of climate change on malaria infections is still uncertain, and the impacts will vary from region to region.

Nevertheless, the close link of the occurrence of malaria and other vectorborne diseases with climatic parameters and the potential changes in the distribution ranges of such diseases warrant responsive health institutions, precautionary action, and monitoring. Climate change-induced droughts, flooding and other extreme weather events degrade and reduce potable water supplies and increase water-associated diseases such as cholera and diarrhea, particularly in areas with inadequate sanitary infrastructures.

Inadequate access to safe drinking water and sanitation, combined with poor hygiene practices, are major causes of ill health and life-threatening disease in developing countries. At present, these diseases already kill an estimated 2.213 million people per year in developing countries, of which about 90 percent are children under the age of five. Women are particularly exposed to water-associated diseases through their traditional chores of washing and water collection.

Migration, and Conflicts

The direct and indirect effects of climate change and their interaction with other vulnerabilities and environmental exposures may lead to mass migrations, as crucial resources become degraded and livelihoods are threatened. Loss of land mass in coastal areas due to sea level rise is, for example, likely to lead to greater permanent or semi-permanent displacement of populations, which may have considerable economic and political ramifications.

Areas most vulnerable to sea level rise lie in the tropics: the west coast of Africa; the north and eastern coast of South America; South and Southeast Asia; and small island states in the Caribbean, Pacific and Indian Oceans. Of the world's 19 mega-cities, 16 are on coastlines and all but 4 are in the developing world. The poor living in Asian mega-cities are particularly at risk, as sea level rise compounds subsidence caused by excessive groundwater extraction in Manila, Bangkok, Shanghai, Dhaka, and Jakarta. To this should be added the risk for potential conflicts, including social unrest, political instability, and wars over decreasing water or other natural resources and possible mass migration due to, for example, land loss or degradation and extreme weather events. Such conflicts may have considerable costs both in macroeconomic terms and in human suffering.

Climate change is expected to have effects on the overall economy of poor countries, thus hampering potential for economic growth. In addition, poor adaptation will increase the impacts of extreme events, increasing the costs of rehabilitation and diverting funds from longerterm development purposes. Current extreme weather events are already taking their toll on developing countries' economies, leading to loss of human and economic capital. Regions where climate change exacerbates climatic extremes and which have limited adaptive capacity will be further constrained in their development prospects due to additional loss of life, private assets, reduced productivity of important economic sectors, and destruction of infrastructure.

This is particularly true for small countries and countries with low economic diversity, where the impact of climatic extremes cannot be well absorbed by economic activity in other regions or sectors. Even though both people and systems appear to be generally more vulnerable to sudden disruptive changes than gradual ones, long-term climate change can be just as harmful. Changes in average climatic conditions, as well as extremes, and loss of productive areas due to sea level rise, have both been highlighted in their projected impacts on the agricultural sector.

Countries where tourism represents a major source of income may be affected by a decrease in revenues due to the effects of both gradual climatic changes and extreme weather events. Such events are likely to alter the attractiveness of certain holiday destinations, for example coral reef mortality is expected to reduce income opportunities for local populations in some regions. All these factors can affect GDP, balance of payments, level of indebtedness, state of public finances, and may divert investments from important development objectives.

Poverty Eradication

Climate change provides an additional threat that adds to, interacts with, and can reinforce existing risks, placing additional strains on the livelihoods and coping strategies of the poor. In 2000, leaders of 189 nations agreed on the Millennium Declaration that outlined eight fundamental goals. Climate change challenges the achievement of the Millennium Development Goals (MDGs) and related national poverty eradication and sustainable development objectives. Unless concrete and urgent steps are undertaken to reduce vulnerability and enhance adaptive capacity of poor people, and unless these actions are integrated in national strategies for poverty eradication and sustainable development, it may be difficult to meet some MDGs by 2015.

Strategies to strengthen capacity to cope with current climate variability and extremes and to adapt to expected future climatic conditions are mutually supportive and will have immediate benefits. They will also help identify and take advantage of the positive impacts of climate change. There is much experience to date of coping with climate variability and disasters from which useful lessons for adaptation can be drawn. Ensuring that the poor are able to adapt to current and imminent climate variability is the first step. The task ahead for the development community is to enhance the adaptive capacity of the poor and poor countries and to help to implement specific actions for addressing climate change impacts.

Natural Disasters and Human Vulnerability

Disasters can occur as a consequence of the impact of a natural or a human-caused hazard. Natural hazards comprise phenomena such as earthquakes, volcanic activity, landslides, tsunamis, tropical cyclones and other severe storms, tornadoes and high winds, river and coastal flooding, wildfires and associated haze, drought, sand and dust storms, and infestations. Human-caused hazards may be intentional, such as the illegal discharge of oil, or accidental such as toxic spills or nuclear meltdown. All of these can expose people, ecosystems, flora and fauna to threats. The poor are the most vulnerable to disasters because they have fewer resources and capacity to prevent or cope with the impacts.

People and the environment are increasingly suffering from the effects of natural disasters. There are a number of reasons for this such as high population growth and density, migration and unplanned urbanization, environmental degradation and possibly global climate change. The sheer scope of the socioeconomic impacts of natural disasters has brought about a shift in the political approach to dealing with the concept of risk in modern societies.

Comparing the past two decades, the number of people killed in natural and non-natural disasters was higher in the 1980s than in the 1990s. However, more people were affected by disasters in the 1990s - up from an average of 147 million a year in the 1980s to 211 million people a year in the 1990s. While the number of geophysical disasters has remained fairly steady, the number of hydrometeorological disasters has increased.

In the 1990s, more than 90 per cent of those killed in natural disasters lost their lives in hydrometeorological events such as droughts, windstorms and floods. While floods accounted for more than two-thirds of people affected by natural disasters, they are less deadly than many other types of disaster, accounting for only 15 per cent of deaths. The social and economic costs of disasters vary widely and are difficult to estimate on a global basis. Insurance claims tend to be misleading as an estimate of the economic impact of disasters.

Considering insured damage claims for the 1999 floods in Austria, Germany and Switzerland, at least 42.5 per cent of damage was covered by disaster insurance. But in Venezuela the same year, only 4 per cent of flood damage was covered. There is a need for reliable and systematic data on disasters to help assess their socio-economic and environmental impacts in both the short and the long term. But although communities in developing countries suffer from numerous local-scale disasters such as wildfires, small floods, droughts and pest infestations, these are often not reflected in disaster statistics.

The most expensive disasters in purely financial and economic terms are floods, earthquakes and windstorms but events such as drought and famine can be more devastating in human terms. While earthquakes accounted for 30 per cent of estimated damage, they caused just 9 per cent of all fatalities due to natural disasters. In contrast, famine killed 42 per cent, but accounted for just 4 per cent of damage over the past decade.

In 1999, global financial losses from natural catastrophic events were estimated to exceed US$100 billion - the second highest figure on record. A total of 707 large events were recorded compared to 530 to 600 events in previous years. It is even more striking that the number of major catastrophic events over the past decade has increased threefold in comparison with the 1960s, while the rate of economic losses has increased by a factor of almost nine over the same period. Between 1995 and 1997, the impacts of natural hazards cost the United States at least US$50 billion a year, or the equivalent of about US$1 billion a week.

The economic losses of the United States because of the 1997-98 El Niño event were estimated at US$1.96

billion or 0.03 per cent of GDP. Ecuador suffered equivalent financial losses but this represented 11.4 per cent of its GDP. The floods in China in 1991, 1994-95 and 1998 caused losses ranging from US$20 to 35 billion. The annual loss from natural disasters over the period from 1989 to 1996 is estimated to range from 3 to 6 per cent of China's GDP, averaging 3.9 per cent. In December 1999, the Anatol, Lothar and Martin storms generated losses in northern Europe amounting to US$5-6 billion.

Less developed countries with limited economic diversity and poor infrastructure must not only rely mostly on external relief if a disaster happens but their economies need more time to recover. In developed economies, governments, communities and individuals have greater capacities to cope with disasters, the economic losses are to some extent absorbed by a diversified economy, and most assets are insured. Among the least developed countries, 24 of the 49 face high levels of disaster risk; at least six of them have been affected by between two and eight major disasters per year in the past 15 years, with long-term consequences for human development.

Since 1991, more than half of all the disasters reported occurred in countries with medium levels of human development. However, two-thirds of those killed came from countries with low levels of human development, while just 2 per cent came from highly developed countries. The effect of development on disasters is dramatic: on average, 22.5 people die per reported disaster in highly developed countries, 145 die per disaster in countries with medium human development, and 1 052 people die per disaster in countries with low levels of development. A number of experts link the current trend in extreme weather events with an increase of the global mean temperature.

Many parts of the world have suffered major heat waves, floods, droughts and other extreme weather events. While individual events, such as El Niño-related phenomena, cannot be directly linked to human-induced climate change, the frequency and magnitude of these types of events are predicted to increase in a warmer world. The changes in the global mean temperature are 'very likely' to affect parameters such as precipitation patterns, wind velocities, soil moisture and vegetation cover which appear to influence the occurrence of storms, hurricanes, floods, drought and landslides. For example, the extent of damage from storm surges can be directly linked to sea level variations.

Climate change and variability alone do not explain the increase in the impacts related to disasters. 'Natural' can be a misleading description for disasters such as the droughts, floods and cyclones which afflict much of the developing world. Identifying humaninduced root causes, and advocating structural and political changes to combat them, is long overdue. For example, destruction of the natural environment because of logging or inappropriate land uses for short-term economic gain is one of the major factors promoting floods or mudslides such as those that hit Venezuela in December 1999.

Similarly, the migration of population to urban and coastal areas increases human vulnerability as population densities increase, infrastructure becomes overloaded, living areas move closer to potentially dangerous industries, and more settlements are built in fragile areas such as floodplains or areas prone to landslides. As a result, natural catastrophes affect more people and economic losses are increased. For example, despite the fact that seismic activity has remained constant over recent years, the effects of earthquakes on the urban population appear to be increasing.

A number of major accidents involving chemicals and radioactive materials have drawn attention worldwide to the dangers of mismanagement, particularly in the transport, chemical and nuclear power sectors. These events often have impacts that transcend national boundaries; they also emphasize the fact that issues of technological safety concern more than just the developed countries.

Some disasters have resulted in the introduction of voluntary or mandatory regulations designed to prevent similar occurrences. Public concern following the explosion in 1976 at a pesticide plant in Seveso, Italy, with the release of -tetrachlorodibenzo-pdioxin (TCDD), led to the introduction in 1982 of a European Directive on the major-accident hazards of certain industrial activities. Similarly, other major accidents, such as the methyl isocyanate leak at Bhopal, India, in 1984 and the Switzerland-Sandoz warehouse fire in Basel in 1989, stimulated legislation in many countries to prevent and control chemical incidents.

Influenced by the Bhopal accident, in particular, the International Labour Office developed in 1993 the Convention Concerning the Prevention of Major Industrial Accidents No. 174 and the Prevention of Major Industrial Accidents Recommendation No. 181. These documents call for an international exchange of relevant information, development of policies aimed at addressing the major accident risks, hazards and their consequences, and recognition that a major accident could have serious impacts on human life and the environment.

Major nuclear accidents such as those at Three Mile Island in the United States in 1979 and at Chernobyl in

1986 have not only catalysed action to strengthen nuclear safety and emergency preparedness but also forced many countries to abandon or severely restrict development of the nuclear power sector. Following the Chernobyl accident, two major international treaties were adopted - the Convention on Assistance in the Case of Nuclear Accident or Radiological Emergency and the Convention on the Early Notification of a Nuclear Accident.

Most recently, the 1994 Convention on Nuclear Safety, committing parties to a higher level of nuclear safety, and the 1997 joint Convention on the Safety of Spent Fuel Management and on the Safety of Radioactive Waste Management were adopted. The 1989 Exxon Valdez oil spill in Alaska resulted in enormous environmental and economic damages and catalysed the development of the 'Valdez Principles' - a voluntary code of conduct for corporate behaviour towards the environment - by the Coalition for Environmentally Responsible Economics (CERES). The 'Valdez Principles' guide corporations in the establishment of environmentally sound policies and require the raising of corporate environmental safety standards and the taking of responsibility for environmental harm that may be caused by them.

Until the 1970s, the international community considered disasters as exceptional circumstances, when local coping capacities were exhausted and external emergency relief was required. The term disaster management was generally equivalent to disaster response and tended to be within the exclusive competence of organizations such as the Red Cross and Red Crescent Societies or national civil defence institutions.

In 1971, the United Nations Disaster Relief Office, now the United Nations Office for the Coordination of

Humanitarian Affairs (UNOCHA), was established to mobilize and coordinate relief activities from all sources in times of disaster. The concept of disaster preparedness was developed during the 1970s and 1980s, and included training and some cross-sectoral activities to increase capacity for rescue, relief and rehabilitation during and after a disaster. But even the most pessimistic forecasts could not have foreseen the upward spiral in negative socio-economic consequences of natural disasters in the closing decades of the 20th century.

The 1990s was declared the International Decade for Natural Disaster Reduction (IDNDR), one of the principal goals of which was to inculcate a culture of disaster prevention through the wider application of known scientific and technological mechanisms by a better-informed population. In the words of United Nations Secretary-General Kofi Annan, 'We must, above all, shift from a culture of reaction to a culture of prevention.

The humanitarian community does a remarkable job in responding to disasters. But the most important task in the medium and long-term is to strengthen and broaden programmes which reduce the number and cost of disasters in the first place. Prevention is not only more humane than cure, it is also much cheaper'. The IDNDR successfully placed risk reduction higher on the political agenda, as well as setting out a number of priorities to be undertaken by countries and regions in the 21st century.

An increasing number of governments and international organizations are promoting risk reduction as the only sustainable solution for reducing the social, economic and environmental impacts of disasters. Risk reduction strategies include:

— vulnerability mapping;

— identification of areas that are safe for settlement and development;

— adoption of building codes based on disaster resilient engineering and on local hazard risk assessments; and

— enforcing these plans and codes by economic and other incentives.

At the global level, the UN has established an International Strategy for Disaster Reduction (ISDR), a global platform aimed at helping all communities to become resilient to the effects of natural disasters and to proceed from protection against hazards to the management of risk through the integration of risk prevention into sustainable development. The strategy - based on the IDNDR experience and developments such as the 1994 Yokohama Strategy and Plan of Action for a Safer World and the 1999 Strategy 'A Safer World in the 21st Century: Disaster and Risk Reduction' - reflects a cross-sectoral and interdisciplinary approach to disaster reduction.

Implementation of the strategy, which is based on the establishment of partnerships between governments, non-governmental organizations, UN agencies, the scientific community and other stakeholders in the disaster reduction community, is an integral part of efforts aimed at promoting the overall goal of sustainable development. It is also an indispensable element in the search for solutions designed to counter the increasing threat posed by natural hazards.

Extreme hydrometeorological events such as flooding and drought are common across Africa, while geophysical events such as earthquakes occur predominantly in

Northern Africa, along the Atlas mountain range, and in the African Rift Valley, which also experiences volcanic activity. The El Niño Southern Oscillation causes significant climatic disturbances in most parts of Africa, either inducing drought or flooding, or increasing sea temperatures leading to cyclones.

These natural events become disasters where large numbers of people or infrastructure are affected, as has occurred over the past 30 years due to high population growth rates, especially in urban centres and drought-prone areas - 34 per cent of Africa's population lives in arid areas compared to just 2 per cent of Europe's population.

Impacts of disasters include loss of lives and livelihoods, damage to infrastructure and communications, interruption of economic activities, and increased risk of disease outbreaks. In many places, these impacts are worsened by poverty and marginalization, and overcrowding. Inadequate, old and deteriorating infrastructure and lack of economic security to provide for times of hardship also compromise people's coping capacities and therefore magnify the impacts of disasters. There is growing concern that the frequency and severity of disasters are increasing at a time when early warning systems are inadequate and disaster management is weak.

Africa has experienced some of the worst droughts and famines in terms of number of people killed or number affected, with particularly severe droughts in 1972-73 and 1984-85, affecting much of Northern, Southern, Eastern and Sahelian Africa. Countries most regularly affected include Botswana, Burkina Faso, Chad, Ethiopia, Kenya, Mauritania and Mozambique, where the impacts of famine are exacerbated by inadequate transport facilities to receive and distribute food and medical aid.

There are some indications that droughts are becoming more prolonged and their impacts more severe. The risk of damage from heavy rain is greater in drier areas than in those that usually receive higher rainfall because there is less vegetation cover to absorb the water and stabilize soils. Expansion of informal settlements into the flood zone is putting many more people at risk of flooding, as illustrated in South Africa's Alexandra Township, Johannesburg, during the floods of 2000 when approximately 3 000 families living in shacks below the floodline were subjected to flooding and outbreaks of cholera.

Disasters can have severe economic impacts which are difficult to calculate. The Western Indian Ocean islands typically experience ten cyclones a year, between November and May, which bring strong winds and heavy rainfall. This causes destruction of infrastructure, particularly in low-lying areas and where settlements have encroached into flood-prone areas. Huge costs are incurred due to destruction of income-generation activities, including tourism revenues, and rehabilitation and replacement of damaged infrastructure and crops.

Globally, Africa suffers the least damage from disasters in purely financial terms but the significance of such losses may actually be greater in terms of impact on economic development. Africa's people and economies are heavily dependent on rainfed agriculture, and are therefore vulnerable to rainfall fluctuations. It is usually the poor who suffer most from flood or drought-induced crop failure, because they often cultivate areas that are climatically marginal for crop production and they cannot accumulate reserves for times of hardship.

Both droughts and floods can result in malnutrition and famine, and the associated food imports and dependency on food aid can affect the economic growth potential of affected countries. In Kenya, low reservoir levels resulting from drought and siltation linked to deforestation led to reductions in hydropower generation, necessitating water and power rationing which devastated the country's economy in 1999 and 2000. Losses from power rationing alone were estimated at US$2 million per day, and the cost of unmet electricity demand was estimated at US$400-630 million, equal to 3.8-6.5 per cent of GDP. In Mozambique, the costs of floods in 2000 were estimated at US$273 million in physical damage, US$247 million in lost production, US$48 million in lost exports and US$31 million in increased imports.

Although climatic variability is a natural phenomenon, the increasing frequency and severity of extreme events can be in part attributed to human activities such as deforestation and inappropriate management of land and water resources. For example, clearing of tropical forests in Central and Western Africa has altered the local climate and rainfall patterns, and increased the risk of drought. Clearing of vegetation may also increase run-off and soil erosion.

Damming of rivers and draining of wetlands reduces the environment's natural ability to absorb excess water, enhancing the impacts of floods. For example, countries in Southern Africa experienced devastating floods in 1999 and 2000 which affected more than 150 000 families. Degradation of wetlands such as the Kafue wetlands in Zambia, damming of rivers, deforestation and overgrazing lowered the environment's ability to absorb excess water and magnified the impact of the floods.

Over the past three decades, millions of Africans have sought refuge from natural and human-made disasters with both environmental and socio-economic impacts. At the end of 2000, there were 3.6 million refugees in Africa, 56 per cent of whom were under 18 years of age. Often refugees are settled in fragile ecosystems where they exert considerable pressure on the natural resources, as they have no other means of survival. Refugee populations also sometimes experience further conflicts with neighbouring communities, through competition for resources.

There have been no concerted regional efforts to manage disasters, and disaster responses in Africa have tended to focus on national and sub-regional levels. Efforts have also concentrated on responses rather than mitigation through improved environmental management and agricultural practices. The unpredictable nature of extreme events, and the weak economic performance of most African countries, makes preparation for, and relief from, disasters all the more difficult.

There have however been some successes in preventing famine resulting from drought, such as the Famine Early Warning System (FEWS) project, the implementation of a new efficient seed distribution system in Niger and promotion of more drought-resistant crop varieties. In Northern Africa, efforts to respond to economic hardship during droughts include financing employment-generating projects to keep farmers from abandoning lands where productivity is declining, while in Eastern Africa afforestation and reforestation projects are being implemented to lessen the impact of future environmental changes, particularly climate change.

In Southern Africa, the SADC Regional Early Warning Unit, the Regional Remote Sensing Project, the Drought

Monitoring Centre and the FEWS Project advise governments on drought preparedness. A drought fund is also in place to mitigate the effects of poor rainfall. In some areas, including parts of West Africa, longterm measures such as urban planning regulations which prohibit developments along watercourses have been promulgated although resource constraints often prevent them from being strictly enforced.

Additional responses include the development and implementation of early warning or forecasting mechanisms such as ENSO forecasting, which has been implemented in Southern Africa and the Western Indian Ocean area. While this has the potential to alert relief organizations and evacuate communities ahead of time, it has been limited by poor communications services. For example, only 152 of every 1 000 people in Africa had radios by 1997.

With global warming, the incidence of drought is likely to increase in many parts of Africa. The frequency and intensity of cyclones and floods in some areas are also likely to increase, adding to the stresses on water and food security, and possibly contributing to outbreaks of disease. For example, Seychelles is currently outside the cyclone zone but sea temperature rise could cause an increase in cyclone intensity and expansion of the cyclone zone to include the islands.

About 75 per cent of the world's major natural catastrophes between 1970 and 1997 occurred in the Asia and the Pacific region, mostly in poverty-ridden developing countries. There has been a general upward trend in the number of natural disasters due to hydrometeorological events in the region, while geophysical disasters such as volcanic eruptions, earthquakes and tsunamis have remained fairly steady.

Vulnerability to disasters is closely linked with population density and economic resources. The impact of natural disasters in the region is severe, with more than 1.4 million people killed, almost 4 000 million affected and US$438 million in damage over the past three decades. During 1991-2000 alone, the total number of deaths caused by natural disasters in the region exceeded 550 000 or 83 per cent of the global total, the majority of them in Asian countries with low or medium levels of human development.

The highest number of deaths occurred in South Asia and the lowest number in Australia and New Zealand, the subregion with the lowest population density and a high per capita income. China experienced more than 300 natural disasters and recorded more than 311 000 deaths during 1971-2000; India with more than 300 disasters suffered more than 120 000 deaths; the Philippines, with nearly 300 events, lost about 34 000 people; Indonesia experienced about 200 disasters with more than 15 000 lives lost; and Bangladesh experienced 181 events and more than 250 000 people killed.

Some areas are more exposed to natural hazards because of location. Cyclones occur most frequently over the Northwest Pacific, over the southern end of the Bay of Bengal, east of India and south of Bangladesh. Bangladesh, China and India are the most flood-prone countries of the region. Hilly and mountainous areas are most prone to landslides, which are aggravated by deforestation and cultivation that destabilizes slopes. Countries along or adjacent to seismic zones are more vulnerable to seismic events, while countries along the Pacific Rim are at risk from volcanic eruptions, particularly Indonesia, Japan and the Philippines. The El Niño phenomenon has significant impacts over large areas in the region, the most affected being Indonesia.

Environmental degradation and change are becoming increasingly important in relation to both the occurrence and impact of natural disasters. Deforestation, for example, is now frequently linked to severe flood events and landslides. Overexploitation of water resources has already resulted in sub-regional environmental disasters such as the drying up of the Aral Sea in Central Asia.

Table 1

Impact of natural disasters in Asia and the Pacific, 1972-2000

	number killed (thousands)	*number affected (thousands)*	*damage (US$1 000)*
South Asia	761	2 164 034	60 881
Southeast Asia	73	284 074	33 570
Northwest Pacific and East Asia	606	1 447 643	317174
Central Asia	3	4 895	986
Australia and New Zealand	1	15 761	21 900
South Pacific	4	4 061	3139
Total	1 447	3 920 467	437 649

Most of the countries in the Northwest Pacific and East Asia sub-region and the Pacific Island countries will be particularly vulnerable to climate change and associated sea-level rise because so many human settlements and so much industrial infrastructure are located in coastal or lowland areas. For the small island developing states, climate change and extreme weather events may also have dramatic impacts on terrestrial biodiversity, subsistence cropping and forest food sources.

Rapid population growth, urbanization and weak land-use planning are some of the reasons why poor people move to fragile and high risk areas which are more exposed to natural hazards. Moreover, the rapid growth of industries in urban areas has induced rural-urban migration. This has sometimes led to more people being exposed to technological hazards such as the disaster in Bhopal, India, in 1984, in which escaping methyl isocyanate from an industrial plant killed more than 3,000 and affected more than 200,000 others.

Asian countries are at different stages of institutional development with respect to disaster reduction. Some, such as Japan, have a long-established system of disaster management. Stimulated by the International Decade for Natural Disaster Reduction (IDNDR), other countries have either strengthened existing frameworks or are formulating new ones. Despite some recent achievements, significant measures and actions still remain to be taken at the regional and national levels to reduce risks and losses due to disasters, namely:

— the impact of environmental degradation needs to be examined - raising awareness about the dangers of environmental degradation among the governments and people is of utmost importance;

— deforestation should be stopped;

— mitigation and preparedness measures already undertaken must be strengthened;

— actions are needed to reduce poverty levels with a view to maintaining the resource base and protecting biodiversity; and

— rural development is a prerequisite for reducing the migration of people to cities and coastal areas.

Disasters resulting from natural and human-caused hazards occur throughout Europe, often leading to significant environmental damage, economic loss, human injuries and premature mortality. The overall consequences depend on both the magnitude of the event and on factors such as population density, disaster prevention, preparedness and response measures and emergency planning. Europe generally suffers less from disasters than many developing countries, due to a higher level of 'coping capacity' in terms of government ability to prepare and respond to disasters.

In Europe, the most common natural disasters are storms and floods, although earthquakes do occur in some countries. Storms and floods are also the most costly in terms of economic and insured losses. The windstorms Lothar and Martin that occurred in December 1999 caused an estimated €5 billion damage to crops, forests and infrastructure while the cost of flood damage between 1991 and 1995 has been estimated at €99 billion.

One of the worst years ever in terms of flood damage was 2000, accounting for almost one-quarter of the total US$10.6 billion insured costs (Swiss Re 2001). In recent years, many European countries have experienced abnormally high precipitation intensity and duration, especially in winter months, which has led to floods in the Czech Republic, France, Germany, Hungary, Italy, Portugal, Switzerland, Ukraine and the United Kingdom. Between 1971 and 1996, 163 major floods occurred in Europe.

The main factors that induce or intensify floods and their impacts include climate change, land sealing, changes in the catchment and floodplain land use, population growth, urbanization and increasing settlement, roads and railways and sometimes hydraulic engineering measures. Forest fires and droughts are a problem in the southern countries along the rim of the Mediterranean and fires are also common in the Siberian region of the Russian Federation where economic recession has caused a severe decline in the response capacity of local authorities and forest fire teams.

Each year, hundreds of thousand of hectares of taiga forest are lost due to fires. Around 80 per cent of forest fires are a result of people's ignorance of fire safety rules. The average annual number of natural disasters appears to be increasing and since the late 1980s there has also been an increase in the impacts of these disasters and their related economic losses - at least in the European Union.

For example, at the French-German border, the floodwaters of the Rhine rose more than 7m above flood level about once every 20 years between 1900 and 1977. Since 1977, that level has been reached on average once every other year. Actions and measures are taken at both national and regional level to reduce the impacts of natural disasters, though there is no targeted policy.

Integrated land-use planning can, to some extent, prevent impacts on humans. Emergency response plans have been produced throughout the European Union to react to various natural disasters, but these appear to be ad hoc, generally not tested, and are considered unlikely to work well in practice. Human-caused disasters cause more human fatalities and economic losses than natural disasters in Europe. Despite overall higher technological and safety levels in Europe, the number of industrial accidents in the European Union continues to rise.

In 1997, there were 37 major industrial accidents - the highest annual number since records began in 1985. In contrast to accidents in fixed installations, major oil spills from marine transport and offshore installation accidents have shown a downward trend although the total number of oil spills appears to be increasing. It is likely that the overall risk from nuclear accidents increased in the 1970s as more plants were commissioned but declined in the 1990s as older plants were taken out of service and the building of new ones either slowed or was completely abandoned due to public pressure. However, quantifying the risk from accidental releases of radionuclides is not possible due to lack of sufficiently detailed, comparable information.

A widespread campaign on increasing the safety of new and already functioning civil nuclear reactors, especially in CEE countries, was catalysed by the 1986 nuclear accident at Chernobyl in the former Soviet Union. Significant resources have been allocated to increasing nuclear safety at nuclear processing plants. However, a complicating factor is the increasing deterioration of the older nuclear power plants in the Russian Federation and Lithuania built to a similar design to the Chernobyl reactor. Analyses of major industrial accidents indicate that component failure and operator error are the two most common immediate causes but the dominant underlying causes identified were poor safety and environmental management.

Lack of expenditure on safety and environmental management, and operating plants past their design life, are often a result of pressure from shareholders wishing to increase profitability, although this may result in major losses in the long run. However, they also reveal gaps in regulation and monitoring. The mining accident at Baia Mare, Romania, in January 2000 served as a rather sobering reminder of the shortcomings in enforcement of environmental regulations in the countries of Eastern Europe.

For many technological disasters, holistic approaches are becoming more prevalent, with increasing attention to reducing the risk of long-term environmental impacts as well as reducing acute health and property damage from accidents. The European Commission's directive on the control of major-accident disasters involving dangerous substances, now also incorporated into the legal systems of most CEE countries, is important in this respect. Its Major Accident Reporting System (MARS) database and the Seveso Plants Information Retrieval System database are practical tools helping countries make decisions related to risk management.

Information on the extent and location of technological hazards is generally improving. Emergency response plans can now be made for technological accidents but further efforts are still necessary to reduce risks. Since pollution does not stop at political boundaries, one of the most important multilateral agreements in this respect is the 1992 Helsinki Convention on the Protection and Use of Transboundary Watercourses and International Lakes, which entered into force in 1996. This convention includes requirements to conduct environmental impact analyses (EIAs), to notify downstream states of accidents and it enforces the 'polluter pays' principle.

The 1991 Convention on Environmental Impact Assessment in the Transboundary Context, which entered into force in 1997, requires parties to notify and consult each other on all major potentially dangerous ongoing projects (ECE 1991). An innovative approach is under consideration with respect to a proposed joint protocol on liability under the Helsinki Convention and the

Convention on the Transboundary Effects of Industrial Accidents. Most European countries are parties to these multilateral treaties, and international cooperation under their provisions helps governments improve national policies in respect of human-caused disaster prevention and mitigation.

In Latin America and the Caribbean, the main natural hazards are droughts, hurricanes, cyclones, tropical storms, floods, tidal waves, avalanches, landslides and mudslides, earthquakes and volcanoes. Mining and oil spill accidents represent the main human-caused disasters in the region. A total of 65 260 deaths due to natural disasters were reported in the region during the 1990s. The deaths were mainly a result of floods (54 per cent), epidemics (18.4 per cent), storms, cyclones and hurricanes (17.7 per cent), earthquakes (5.2 per cent) and landslides (3.2 per cent).

Considering that floods and landslides are often associated with storms and hurricanes, this means that three-quarters of total human losses due to natural disasters in the region have a hydrometeorological origin. The number of deaths due to disasters declined markedly between 1972 and 1999, coinciding with the global trend. Total fatalities in the 1990s were less than one-third of those in the 1970s while the number of people injured fell by almost one-half. This trend can be explained by fewer severe earthquakes in densely populated or highly vulnerable areas and by the establishment of early warning systems and disaster-preparedness measures in some countries over the past 30 years.

Economic losses caused by disasters increased by almost 230 per cent between the 1960s and the 1990s, again reflecting a global trend. The best known hydrometeorological event is the El Niño phenomenon, the impacts of which can be severe. For example, after the El Niño of 1983, Peru's GDP fell by 12 per cent, mostly because of a reduction in agricultural output and fishery. The national economy took a decade to recover. Damage in the Andean Community countries (Bolivia, Colombia, Ecuador, Peru and Venezuela) due to the 1997/98 El Niño was estimated at more than US$7 500 million.

Most countries in Central America and the Caribbean are within the hurricane belt, on both the Atlantic and Pacific coasts. Hurricane Mitch, which struck the region in 1998, affecting mostly Honduras and Nicaragua, killed more than 17 000 people and left three million homeless with damage estimated at US$3 000 million. The hurricane also caused fatalities and serious environmental and economic damage in Costa Rica, Dominican Republic, El Salvador and Guatemala.

The 1999 floods on the northern coast of Venezuela also had a strong impact, with damage estimated at more than US$ 3 200 million or 3.3 per cent of the country's GDP. In the state of Vargas, the hardest-hit area, more than 230 000 jobs were lost. The state of Miranda was also badly hit: the El Guapo dam collapsed, causing water shortages and 60 per cent of crops were reported lost. It is estimated that there were 30 000 deaths, 30 000 families left homeless and more than 81 000 dwellings destroyed.

Seismic and tectonic activities are particularly intense along the Pacific Ocean coast and in the Caribbean basin due to pressures generated between oceanic and continental plates. Such activity creates a relatively high risk of earthquakes, tsunamis and volcanic eruptions which in some areas add to the already high risk of hurricanes and floods. Between 1972 and 1999, extreme geological events killed 65 503 people and affected 4.4 million others.

3,000 kg of mercury have been dumped and a spill of 1.5 million litres of cyanide-polluted waste has been reported in the Omai and Esequibo rivers in neighbouring Guyana.

Many countries, especially those located on islands, are vulnerable to natural disasters. The major concerns with regard to policy include the following:

- deficiencies in disaster prevention, including the lack of zoning of vulnerable areas during the development planning process;
- weak mitigation mechanisms;
- deficiencies and limited use of anti-seismic building measures, as well as inadequate administrative arrangements and human resources for enforcement;
- lack of insurance policies for low-income households; and
- inadequate support systems for affected communities.

Improving management is critical to disaster reduction, especially non-structural mitigation actions using natural mechanisms. For example, wetlands reduce floods, woodlands reduce landslides and mangroves lessen the effect of coastal storms and extreme tides. In general, good land use maintains healthy ecosystems, provides resources and facilitates non-structural mitigation action. This strategy is particularly attractive in countries where risk insurance and structural mitigation come at a high price.

Given the enormous economic, social and environmental burden of disasters, considerable attention has been paid during the past decade to disaster preparedness, assessment and mitigation. Many of the actions took place in the context of the International

Decade for Natural Disaster Reduction (IDNDR). At the regional level, its mandate for promoting international cooperation in this field was supported by the Inter-American Conference on Natural Disaster Reduction held in Cartagena, in March 1994.

Several countries in the region - such as Brazil, Costa Rica, Cuba, Chile, Colombia, Guatemala, Nicaragua and Panama - have created and strengthened national institutional frameworks in the area of disaster management. These include the Centre for Coordination of the Prevention of Natural Disasters in Central America, established in 1988, and the Caribbean Disaster Emergency Response Agency, established in 1991. Under the auspices of the Organization of American States, the Inter-American Convention to Facilitate Disaster Assistance was adopted in 1991 and entered into force in 1996.

Experience has shown the positive effects of planning and building institutional capacities. A fundamental element is to strengthen and standardize data production methods at a regional level, not only to prevent inconsistencies during emergencies but also to assess losses. Also important are efforts to identify the vulnerability of the territories and populations when faced with natural and human-made hazards. The prevailing disaster response is directed towards risk management. It has a growing component of local and community participation, and makes non-centralized use of non-governmental organizations and citizen groups. Within this framework, a new vision is emerging: the development process must reduce risk by lessening populations' and territories' social, economic and environmental vulnerability.

Natural hazards such as earthquakes, volcanic eruptions, tornadoes, hurricanes, ice storms, droughts, dust storms and other extreme events threaten different parts of North America. Floods and forest fires are also priority concerns. North American governments have put in place many response mechanisms to prevent and alleviate the harm caused by such factors. Despite strong regulations governing the handling of hazardous material, occasionally serious accidents occur, prompting further preventive legislation.

Floods and Climate Change

The disruption and intensification of the Earth's water cycle is believed to be one of the most fundamental effects of climate change. Changes may already be occurring in North America's hydrological conditions, as demonstrated by the increase in average annual precipitation over the past 30 years. In the United States, the average amount of moisture in the atmosphere increased by 5 per cent per decade between 1973 and 1993. Most of the increase has been due to heavier precipitation events resulting in floods and storms.

During the 1960s and 1970s, more than 90 per cent of the natural disasters in the United States were the result of weather or climate extremes. Flooding is natural and essential to the health of watersheds but floods can also be destructive and cause economic damage. In response to these events, the United States introduced the National Flood Insurance Act of 1968 and the 1974 Disaster Relief Act. Many of the separate and fragmented responsibilities of parallel state and local level disaster programmes were merged in 1979 under the Federal Emergency Management Agency.

In 1975, Canada introduced the Flood Damage Reduction Program (FDRP) and, in 1988, it established Emergency Preparedness Canada (EPC). These programmes provided better flood mitigation, preparation, response and recovery measures. Evidence shows that deaths and damage from floods have increased sharply since the early 1970s. More people and their settlements are exposed to floods because of population increase and concentration, and increasing affluence.

A tendency to settle in flood-prone areas is also influenced by a perception that risk has been lowered by protective structures such as dams, dykes and diversions, and because of the availability of disaster relief. Structures that prevent rivers from flooding often provoke extremely damaging floods when water eventually overflows. In the 1990s, the United States, which is subject to more frequent and severe weather events than Canada, began to encourage non-structural approaches to flood prevention such as resettlement projects and wetland restoration.

In Canada, settlement in flood-prone areas has been discouraged through mapping and the designation of more than 320 flood risk areas. Canada established the Office of Critical Infrastructure and Emergency Preparedness (OCIPEP) in 2001 to develop and implement a more comprehensive approach to disaster prevention. According to some climate change models, the magnitude, frequency and cost of extreme hydrological events in some regions of North America are forecast to increase. Projected effects of climate change include changes in the El Niño.

An uncommonly strong El Niño in 1997-98 is believed to account for heavy floods in Florida, California, some midwest states and parts of New England. Where rainstorms intensify and flooding increases, there is greater potential for damage to lowlying settlements and dock and port facilities as well as for problems with water distribution and sewage systems that can have health implications. The International Joint Commission assists both governments in managing their shared waters. In a report on the 1997 Red River flooding, it cautioned that, given impending increased flooding due to climate change, a comprehensive, binational strategy should be developed and implemented.

Forest fires are a natural part of North America's landscape and play an important role in maintaining and regenerating some types of forests. Wild fires ignited by lightning are useful for clearing out old and dead trees which are then quickly replaced by robust new trees. Such fires open spaces for new seedlings, help increase diversity, clear debris and increase the availability of nutrients.

Since the 1970s, the annual area burned by forest fires has grown. The increase has been due to a number of factors: fuel build-up from past fire protection programmes; changes in policy related to prescribed burning; and increased public access to forests. Climate change has also been implicated. The relative importance of these factors is controversial. The United States has long had an aggressive policy of fire suppression and, by the 1970s, fires were kept to about 2 million ha a year in the lower 48 states compared to the 16 million ha burned every year in the 1930s.

As a result, species normally eliminated by fire became dominant. Dead trees accumulated during periods of drought, creating excessive fuel loads. Fire suppression prevented natural low-intensity fires from burning this accumulated fuel. The result was increasingly large and disastrous fires. The importance of periodic natural fires began to be recognized in the 1970s. US policies of suppressing all fires before they reached 4 ha in size by 10 am the next day ceased in the late 1970s. Decisions were taken not to interfere with fires in wilderness areas or national parks unless people or neighbouring land was threatened. Such fires are either purposefully lit or are lightning fires that are allowed to burn. Annually, more than 2 million ha are treated by prescribed fire in the United States.

These policies have not been without controversy, however. In 1988, parts of Yellowstone - the largest National Park in the United States - were allowed to burn after being struck by lightning. The fires spread quickly because of severe summer drought and high winds. Eventually a decision was made to suppress the fires. At the cost of US$120 million, this was the costliest fire-fighting event in US history. The challenge of managing wildfires has been exacerbated by population increases close to fireprone areas. It is estimated that in the 1990s wildfires damaged six times as many homes as in the previous decade.

Wildfires also create smoke hazards and some highways, airports and recreation areas periodically have to close because of reduced visibility. Smoke also constitutes a health hazard, due to the toxic chemicals it contains. Changes in climate that may bring drier conditions and more severe storms may also play a role in changing fire patterns. In 1989, for example, record fires burned in western Canada and the areas east of James Bay. They were caused by unusual weather conditions and an unprecedented heat wave in the Arctic. The severity of Canada's 1995 fire season, which burned 6.6 million ha of forestland, was also due in part to extremely dry conditions. In the future, North America's annual fire severity rating may well increase due to climate change, which is predicted to increase the number of lightning strikes and the intensity and frequency of windstorms.

West Asia is arid and vulnerable to drought, with rainfall scanty and variable. Nearly 80 per cent of the region is classified as semi-desert or desert. Drought is the most important natural disaster in the region. Rainfall appears to be declining in some countries bordering the Mediterranean Sea. Over the past 100 years, precipitation has decreased by more than 5 per cent over much of the land bordering the Mediterranean with a few exceptions such as Libya and Tunisia. The region suffered droughts during the 1930s, 1960s and the 1990s. In the winters of 1991-92 and 1992-93, snowfall was rare in many areas of the eastern Mediterranean.

Cycles of drought have become intense and more frequent. The 1998-99 drought affected many countries and Syria was the worst hit, suffering its worst drought in 25 years. The most direct effects of the drought were crop failures and a decline in cereal and livestock production. In Iraq, for example, cereal production declined by 20 per cent compared to the previous year and by 40 per cent compared to the average production for the previous five years.

A report by a FAO/WFP mission to Syria stated that a large proportion of the nomadic herders were facing 'financial ruin', with 4,700 households seriously vulnerable to food shortages and in urgent need of food assistance. Cereal production was also seriously affected. Barley harvest was estimated to be only 380,000 tonnes - less than half the 1998 total and down 72 per cent from the previous five-year average. Local needs had to be satisfied through imports. Reduction in wheat production was less severe because 40 per cent of Syria's wheat fields are irrigated. Jordan was also adversely affected by the drought, which reduced the country's wheat and barley production in 1999 by 88 per cent.

Drought results in economic, social and environmental problems. Economic hardships during drought intensify and can lead to social conflict between land users, especially in the Mashriq countries and in Yemen where an agricultural economy prevails. Drought is also a major limiting factor to the region's economic development, affecting the development of agricultural and water schemes, and ultimately food production. Forage and fodder become scarce in rangelands during droughts. In addition, the decline in cereal production and the limited availability of crop residues worsen the impact of drought on the sheep population and consequently on human well-being.

Loss of sheep and the high price of supplementary feed led to a significant drop in farmers' incomes and many families were forced to sell off their animals and other assets at low prices. Land degradation, mostly in the form of desertification, is one of the region's most serious problems. Although desertification is often attributed to poor land use practices, drought deepens the effect and extends the area prone to desertification to encompass areas normally not at risk. Decreases in plant cover due to drought may also increase erosion and lead to a nearly irreversible loss of productive potential and subsequently desertification.

Nations have responded to drought by improving national efforts to combat desertification and joining international ones with the same aim such as the United Nations Convention to Combat Desertification. Under the auspices of this international treaty, national action programmes have been developed and a subregional action programme to combat desertification and drought was adopted in 2000. Human-induced disasters are mostly associated with the oil industry. Intensive oil extraction in the region results in frequent oil discharges into the Gulf. It is estimated that about 10 per cent of the oil discharged in the region enters the marine environment.

Accidental oil spills also occur, with three such accidents being among the world's 20 largest: 300 million litres from the Nowruz Platform on 26 January 1991, 144 million litres from the tanker Sea Star on 19 December 1972 and 118 million litres from storage tanks in Kuwait on 20 August 1981. However, the biggest oil spill took place in January-February 1991, during the 1990-91 Gulf War, when 9 500 million litres of oil were deliberately released in the desert. An estimated 1 500 million litres of oil were released into Gulf waters and more than 600 Kuwaiti oil wells were set on fire. This humanmade disaster had enormous impacts on the environment and human health.

The long-term environmental effects of the Gulf War may persist for decades. Besides land and marine pollution, huge quantities of pollutants such as sulphur dioxide, nitrogen oxides, carbon monoxide and particulate matter were emitted by the burning oil wells. The high particulate matter levels have been associated with an allergic response in people. Hospital studies indicate that about 18 per cent of Kuwait's civilian population suffer from some respiratory complaint, primarily asthma, compared to roughly 6 per cent in the United States.

Along with natural disasters, the region has been plagued with wars. Since the beginning of the 20th century, the region has witnessed the 1948 Arab-Israeli war, the 1967 Six-Day war, the 1973 October war and the Israeli invasion of Southern Lebanon in 1982. In the 1980s and 1990s, the first and second Gulf wars caused major environmental problems. Environmental pollution was a major impact. Fires were set deliberately in forests, and water resources were polluted and/or destroyed. Artillery fire destroyed land resources.

Marine resources were polluted as well as the atmosphere from oil well fires and soils were contaminated by oil spills during the second Gulf War. Wars create refugees. In the aftermath of the 1948 Arab-Israeli war, more than 750,000 Palestinians were left landless and homeless. A second wave of approximately 350,000 Palestinians and more than 150 000 Syrians became refugees at the end of the Six-Day War. Towns and villages in Palestine and Golan Heights were depopulated and destroyed.

Today, there are about 3.8 million refugees in 59 camps registered with the United Nations Relief and Works Agency. Palestinian refugees are scattered in a number of countries, including Jordan, Lebanon and Syria. Most live in poor conditions, putting additional stress on already limited natural resources.

Impacts of natural hazards, combined with extreme polar climatic conditions, and vulnerable ecosystems and infrastructure can easily result in disasters in the Arctic. For example, during the five-year period 1996-2001, there were two catastrophic floods in the Lena River that exceeded all previous records. In the winter of 2001, temperatures hit a record low, some rivers froze solid, and therefore took longer to thaw, and blocks of ice clogged the natural flow. In addition, in that year, the snowfall was particularly severe. The water levels in the central part of the Lena exceeded the normal average by 9 metres or more.

Economic losses and environmental devastation were severe. Because climate change is likely to increase precipitation in the catchment areas of Arctic rivers, there may be a corresponding increase in the frequency and magnitude of floods. The temperature increase observed over the Arctic land masses in recent years results in permafrost thawing in many areas. In the developed areas of the Arctic, efforts will be needed to reduce the impacts of thawing on buildings and transport infrastructure.

Another natural disaster affecting the Arctic ecosystem is pest invasion, which can devastate a forested area and affect the related economic activities. Pest outbreaks are a major problem in the forest-tundra zone. The spruce bark beetle has caused serious destruction and forest death in the spruce forests of Alaska. In Scandinavia, autumn moths cause massive defoliation of birch forests at about 10-year intervals. These forests do not recover for up to centuries because of the slow recovery rate of vegetation in the Arctic.

With the exception of Finland, all the countries bordering the Arctic area have oil terminals, or major transportation routes of oil or hazardous materials in their Arctic areas. Other human activities include the exploitation of petroleum and mineral resources by all countries except Finland and Sweden. Iceland has a hazardous materials waste site, and the Russian Federation has several nuclear sites and radioactive waste sites in its Arctic area.

An environmental risk survey of human activities in the Arctic, carried out under the auspices of the Arctic

Council, concluded that the greatest threat from a release of a pollutant requiring emergency response is the transportation and storage of oil. Nuclear sites, although assessed as less of a threat overall, could affect much larger areas. Pipeline ruptures and leakages, such as in the Usinsk area of Russia in 1994 when 116 million litres of crude oil were spilled, and the Exxon Valdez tanker accident in Alaska in 1989 with almost 50 million litres of crude oil spilled, are examples of catastrophic environmental impacts in the region.

Many smaller accidents, such as uncontrolled oil gushers and the accidental discharge of contaminated mud during drilling, also result in local environmental pollution. Both past and current activities involving radioactive materials in the Arctic create a high potential risk of accidents, although there has been no large-scale radioactive pollution yet. For example, accidents such as the sinking of the Soviet nuclear submarine Komsomolets in 1989 and the Russian nuclear submarine Kursk in 2000, and the crash of a US aircraft with nuclear weapons aboard near Thule in Greenland in 1968, did not result in the release of radioactive substances to the environment.

The Soviet Union dumped high, intermediate and low level radioactive waste in the Kara and Barents Seas between 1959 and 1991, including six nuclear submarine reactors and a shielding assembly from an icebreaker reactor containing spent fuel. Since then, the research and data collected have indicated that no significant amounts of radioactive materials have migrated from the dumping, and only very local samples show elevated radionuclide levels. The major risks may be over the long term as the containers corrode.

Radioactive contamination from European reprocessing plants in the 1970s and atmospheric weapons testing in the 1960s have contributed to current low-level Arctic contamination. Governments, businesses and international organizations are all taking action to increase disaster preparedness in the region. Intergovernmental cooperation is carried out on both a bilateral and a multilateral basis, especially via the Arctic Council. Two of the Arctic Council's programmes - Emergency Prevention, Preparedness and Response (EPPR), and Protection of the Arctic Marine Environment (PAME) - have produced important information and guidelines on environmental risks in the Arctic.

For example, EPPR developed the Arctic Offshore Oil and Gas Guidelines aimed at regulatory agencies in 1997. A guideline on the transfer of petroleum products from ship to shore and ship to ship has been produced by PAME. The IUCN and the Oil and Gas Producers Association have prepared guidelines for environmental protection in the Arctic and sub-Arctic. Uzbekistan, Kazakhstan, and other Central Asian states used this water to grow cotton and other export crops, in the face of widespread environmental consequences, including fisheries loss, water and soil contamination, and dangerous levels of polluted airborne sediments.

The Aral Sea is one of the greatest environmental catastrophes ever recorded. Humans have made use of the waters of the Aral basin for thousands of years, borrowing from its two major rivers: the Amu Darya, which flows into the Aral Sea from the south; and the Syr Darya, which reaches the sea at its north end. The Kara Kum Canal opened in 1956, diverting large amounts of water from the Amu Darya into the desert of Turkmenistan, and millions of hectares of land came under irrigation after 1960.

While the sea had been receiving about 50 km3 of water per year in 1965, by the early 1980s this had fallen to zero. As the Aral shrank, its salinity increased, and by the early 1980s commercially useful fish had been eliminated, shutting down an industry that had employed 60 000. The declining sea level lowered the water table in the region, destroying many oases near its shores.

Over-irrigation caused salt build-up in many agricultural areas. By the beginning of the 1990s, the surface area of the Aral had shrunk by nearly half, and its volume was down by 75 per cent. Winds picked up sediments laced with salts and pesticides, with devastating health consequences for surrounding regions.

Reducing Vulnerability

There is a large and widening vulnerability gap between well-off people, with better all-round coping capacity, who are becoming gradually less vulnerable, and the poor who grow increasingly so. It is vital to the sustainable development effort that this gap is addressed, as well as vulnerability itself. For the most significant improvements, priority should go to policies that reduce the vulnerability of the poor as part of general strategies for poverty reduction. This is in keeping with the general priority being given to poverty reduction as essential to sustainable development. Increasing human vulnerability is only now achieving wide recognition, so that few existing policies specifically address this issue.

However, a number of studies, programmes and projects are currently addressing aspects of human vulnerability and these have already yielded valuable lessons for future policy action. Two types of policy response are possible: reducing the threat through prevention and preparedness initiatives, and improving the coping capacity of vulnerable groups to enable them to deal with the threat.

Exposure to threats can be minimized by reducing the risk - in other words the probability that damage will occur. In theory, exposure can always be avoided by moving people out of hazardous situations but this is not always feasible in practice. Currently, the prediction of threats is an imperfect science. People will not evacuate their homes and businesses unless absolutely necessary, and a few false alarms will quickly discourage any further response.

Human exposure can be reduced by reinforcing infrastructure - for example, by upgrading building codes, improving flood control, planting trees for soil stabilization or avalanche control, and providing safe havens or shelters. Many of these measures require significant long-term investments. Better environmental management, improved policies to protect ecosystems and environmental restoration can be effective and practical ways to reduce vulnerability.

In the long term, every effort to achieve sustainability in natural resource use, to reduce waste generation and pollution, and to bring society back into balance with the local environment and global systems should reduce human vulnerability. One of the main goals of integrated environmental planning should be the integration of vulnerability assessment and reduction. Many natural systems evolved in response to specific major environmental threats and have a builtin capacity to absorb them.

Vegetation stabilizes stream banks, slows run-off and prevents erosion. Beaches absorb wave energy and protect coastlines. Environmental threats may well increase with the destruction of such natural defences. Their restoration is the best response to the problem because it is usually less expensive and more permanent than artificial defensive construction which sometimes simply moves the threat elsewhere. Many flood control works have aggravated problems elsewhere on river systems and are now being systematically reversed.

Improving the coping capacity of groups at greatest risk can do much to reduce the damage caused by extreme events or environmental degradation. The ability to cope with threats includes the ability to absorb impacts by guarding against or adapting to them. It also includes provisions made in advance to pay for potential damage, for instance by mobilizing insurance repayments, savings or contingency reserves. People can draw on both tangible and hidden assets for coping with change, assets that can help to reduce the probability and magnitude of harm. They can be helped to identify and mobilize whatever assets they have in time of need and these may be decisive factors in damage prevention.

They include economic assets, social and political assets, ecological assets, infrastructure assets and personal assets. Strategies that take account of the existing assets of vulnerable groups and their likely needs may also cushion the damaging impacts of unavoidable events or catastrophes. Restitution of lost or damaged assets through rapid response to sudden or extreme events in the form of rescue, relief and rehabilitation (for instance, provision of clean water, health care, shelter and food) may be all that is required to reduce hardship to manageable levels.

Institutional arrangements - including the preparedness of public, private and social services -are an important aspect of coping.

Institutional preparedness can be a decisive factor in reducing vulnerability. For example, the flooding of the River Oder in 1997 caused less damage on the German side than it did in Poland because the Germans were better prepared. People in vulnerable areas should make institutional arrangements to respond to potential crises. This requires foresight but often can be done at little or no cost. The UNEP Awareness and Preparedness for Emergencies at Local Level programme (APELL) is a good example of institutional preparation to cope with potential environmental threats. Where a threat cannot be reduced or eliminated, adapting to it can be an effective response.

Adaptation refers both to physical adjustments or technical measures and changing behaviour, economic activities and social organization to be more compatible with existing or emerging conditions or threats. The latter requires adaptive capacity, including the ability to develop new options and to deliver them to vulnerable populations. Some environmental changes, such as expected climate change from global warming, have such long lead times that some degree of environmental change is inevitable even if measures to control the situation are implemented rapidly.

Some adaptation measures may then be essential. Efforts to predict the probable impacts of climate change should help to determine the adaptive actions that are necessary and the speed with which they should be implemented. Various investments in adaptive capacity have been made following advances in early warning. Several countries have tried to change patterns of agricultural practice so that crops more suited to periodic changes in growing conditions can be grown in years affected by climate fluctuations associated with El Niño and La Niña events. The risk of crop failure is thus reduced.

One of the most effective responses to human vulnerability to environmental change is to strengthen mechanisms for early warning. Many actions can be taken to protect life and property if warning is received in time. While some threats are inherently unpredictable, many of those arising from threats from environmental degradation and mismanagement, and from human activities, can now be anticipated with some precision.

Early warning capacities are increasing steadily with technological advances in environmental observing, assessment and communications. Examples are the cyclone early warning systems that have been established in India and Mauritius. Conventionally, early warning means an urgent indication of an impending hazard. There is a need for both sudden onset warnings, for imminent disaster threats such as tropical storms and floods, and slow onset warnings for disseminating information about disasters that may develop over time such as famine and drought.

The term early warning is often taken to mean 'prediction' when in fact the occurrence of many threatening events is essentially unpredictable. Early warning simply means that an event is imminent and the time to escape from it or take action against it is now. Early warning information can be produced in the context of a broader vulnerability assessment process, which includes the production and communication of forecast information and the incorporation of that information in user decisions. To be effective, an early warning system must be able to stimulate a timely response before an event takes place.

It must identify who are the users of early warning information and what is the most efficient way to reach them with credible information to enhance their powers of decision-making. It must then translate relevant data into early warning indicators that decision-makers can easily interpret and use. Ultimately, the single most important factor that will lead governments to incorporate the use of early warning systems and information in decision-making is the political will to invest in response systems, both nationally and internationally. One example of an operational early warning system which has generated such a response is the Famine Early Warning System Network for Africa.

Assessment Measures of Vulnerability

Vulnerability assessment measures the seriousness of potential threats on the basis of known hazards and the level of vulnerability of societies and individuals. can be used to translate early warning information into preventive action and is a necessary element in early warning and emergency preparedness. Ideally, the results should be incorporated directly into the long-term planning of institutions and governments, and should foster institutional responsiveness to increasing vulnerability, and action for disaster preparedness and mitigation.

Vulnerability assessments are widely applied and used in the fields of climate change and natural disaster management, where they provide the basis for effective warning systems. Assessments of vulnerability can be made for both people and the environmental systems that provide goods and services. They should identify the location of vulnerable populations, the threats to their wellbeing and the extent of their vulnerability; the risks to the environmental capacity to provide goods and services; and the preventive steps that can be taken improve environmental conditions and reduce the negative impacts of human action on the environment.

This information is then assembled into a knowledge base that is accessible, scientifically reliable and easy to use, which can help policy-makers and planners seeking to formulate adequate responses. Calculation of coping capacity would be a valuable tool for understanding how and why burdens of environmental degradation are unevenly distributed around the globe, and why the potential impact of different threats may be more or less catastrophic depending upon a group's ability to cope.

In the case of diseases such as cholera, governments of highincome countries would be likely to respond to the risk of an outbreak with costly prevention and early warning programmes such as a cholera-monitoring network. Yet this response would not be affordable in many other parts of the world. When calculating vulnerability, geographical scale is important. A single national figure may hide many significant variations. Although an assessment for high-income countries would show low overall vulnerability, there may be sub-populations that are highly vulnerable.

For instance, one country may be less vulnerable to outbreaks of vector-borne disease triggered by climate change than another, because of the medical system's capacity to respond, but those without medical insurance may still be particularly vulnerable. Furthermore, societies that are well equipped to cope with present vulnerability may lack the experience or technology to respond to emerging threats.

4

Environment and Sustainable Development

The environment has always been critical to life but concerns over the balance between human life and the environment assumed international dimensions only during the 1950s. At the end of the 1960s, the voice of environmental concern was heard almost uniquely in the West. In the communist world, the relentless destruction of the environment in the name of industrialization continued unabated. In developing countries, environmental concerns were regarded as Western luxuries.

'Poverty is the worst form of pollution,' held India's Prime Minister, Indira Ghandi, who played a key role in orienting the agenda of the UN Conference on the Human Environment, held in Stockholm in 1972, towards the concerns of the developing countries. In the early 1970s, attention was focused first on the biophysical environment, for example, on issues of wildlife management, soil conservation, water pollution, land degradation and desertification - and people were considered as the root cause of such problems. In the West, there were two principal schools of thought about the causes of environmental degradation: one school blamed greed and the relentless pursuit of economic growth; the other blamed population growth.

The Foundation of Modern Environmentalism

The world of 1972 was very different from that of today. The Cold War still divided many of the world's most industrialized nations, the period of colonization had not yet ended and, although e-mail had just been invented, it was to be more than two decades before its use became widespread. The personal computer did not exist, global warming had only just been mentioned for the first time, and the threat to the ozone layer was seen as coming mainly from a large fleet of supersonic airliners that was never to materialize.

Although transnational corporations existed and were becoming increasingly powerful, the concept of globalization was still 20 years away. In South Africa, apartheid still held sway and in Europe the Berlin Wall stood firm. The world of the early 1970s was thus fiercely polarized, and in many different ways. Against this backdrop, it was surprising that the idea of an international conference on the environment should even be broached; it was even more surprising that one should actually take place; and it was astonishing that such a conference could give rise to what later became known as the 'Stockholm spirit of compromise' in which representatives of developed and developing countries found ways of accommodating each other's strongly divergent views. The conference was hosted by Sweden following severe damage to thousands of Sweden's lakes from acid rain falling as a result of severe air pollution in Western Europe.

The United Nations Conference on the Human Environment, held in June 1972, was the event that turned the environment into a major issue at the international level. The conference drew together both developed and developing countries, but the former Soviet Union and most of its allies did not attend. The Stockholm Conference produced a Declaration of 26 Principles and an Action Plan of 109 recommendations. A few specific targets were set - a 10-year moratorium on commercial whaling, prevention of deliberate oil discharges at sea by 1975 and a report by 1975 on energy uses.

The Stockholm Declaration on the Human Environment and Principles constituted the first body of 'soft law' in international environmental affairs. The conference also established the United Nations Environment Programme as 'the environmental conscience of the UN system'. It is easy to claim that many of the major environmental milestones of the 1970s followed directly from Stockholm. It is important to remember, however, that Stockholm was itself a reflection of the mood of the times, or at least of the views of many in the West.

One area in which governments and other stakeholders recorded qualified successes in the 1970s was wildlife conservation. This was achieved through a combination of legal actions at the global level which were (and still are) enforced at the national level with variable effectiveness. The foundation for some of these successes was laid by multilateral environmental agreements such as the:

— 1971 Convention on Wetlands of International Importance Especially as Waterfowl Habitat (Ramsar);

— 1972 Convention Concerning the Protection of the World Cultural and Natural Heritage (World Heritage);

— 1973 Convention on International Trade in Endangered Species of Wild Fauna and Flora (CITES); and

— 1979 Convention on the Conservation of Migratory Species of Wild Animals (CMS).

The Ramsar Convention

The Ramsar Convention predates the Stockholm Conference, having been opened for signature in 1971. The convention, which became effective two years after Stockholm, had 130 parties as of December 2001. It was developed largely out of NGO-led activities in the 1960s concerned with bird life and habitat. While its main focus was initially the conservation of waterfowl and their habitats, it now also deals with water quality, food production, general biodiversity and all wetland areas, including saltwater coasts.

Parties are obliged to list at least one wetlands site of importance, establish nature reserves, make wise use of those sites, encourage the increase of waterfowl populations on appropriate wetlands, and supply information on implementation of policies related to the sites. More than 1,100 areas, covering 87.7 million ha, are currently designated Ramsar sites, enhancing wildlife conservation in different regions.

The World Heritage Convention

The World Heritage Convention, negotiated in 1972, is administered by the United Nations Educational, Scientific and Cultural Organization (UNESCO). It had 161 parties in mid-2001. Since 1972 when the Galapagos Islands were put under the aegis of UNESCO as 'a natural university of unique species', a total of 144 sites in different regions were designated as natural heritage sites as of December 2001. An additional 23 sites were of both natural and cultural significance. The impact has been greater awareness of the importance of these sites for both present and future generations. However, the early 2001 oil spill adjacent to the Galapagos Islands, which threatened species and habitats, underlines the fact that environmental management systems may never be foolproof.

CITES

At the time of Stockholm, it was reported that 150 species of birds and animals had already been 'exterminated' and about 1 000 more were threatened with extinction (Commission to Study the Organization of Peace 1972). A UN Commission recommended the identification of endangered species without further delay, the conclusion of appropriate agreements and establishment of institutions to spearhead wildlife conservation, and the regulation of the international trade in threatened species.

The Commission's recommendation virtually endorsed a 1963 resolution by members of the World Conservation Union (IUCN) which catalysed the drafting of the CITES convention. The convention was eventually adopted in 1973 and became effective two years later. The convention controls and/or bans international trade in endangered species, including about 5 000 animal and 25 000 plant species. Controversy over charismatic species such the African elephant and the whale have often overshadowed the attention that has been placed on other species.

In terms of demonstrable action, Stockholm apparently achieved much. While many of its 109 recommendations remain unfulfilled, they serve -now as then - as important targets. Equally important, however, were the Conference's achievements in repairing rifts, and in narrowing the gap between the views of the developed and the developing nations. The first attempt at this had been made at a conference in Founex, Switzerland, in 1969, and the Founex Report of June 1971 identified development and environment as 'two sides of the same coin'.

The Drafting and Planning Committee for the Stockholm conference noted in its report in April 1972 that 'environmental protection must not be an excuse for slowing down the economic progress of emerging countries'. Further progress had to wait until 1974 when a symposium of experts chaired by the late Barbara Ward, was held in Cocoyoc, Mexico. Organized by UNEP and the United Nations Commission on Trade and Development (UNCTAD), the symposium identified the economic and social factors which lead to environmental deterioration.

The Cocoyoc Declaration - the formal statement issued by the symposium - was influential in changing the attitudes of leading environmental thinkers. What was said at Cocoyoc foreshadowed the first paragraph of the World Conservation Strategy published in 1980 and was re-stated in GEO-2000 in 1999:

'The combined destructive impacts of a poor majority struggling to stay alive and an affluent minority consuming most of the world's resources are undermining the very means by which all people can survive and flourish'. Other statements in the Cocoyoc Declaration illustrate awareness of the difficulty of meeting human needs sustainably from an environment under pressure:

— 'The problem today is not one primarily of absolute physical shortage but of economic and social maldistribution and usage.'

— 'The task of statesmanship is to guide the nations towards a new system more capable of meeting the inner limits of basic human needs for all the world's people and of doing so without violating the outer limits of the planet's resources and environment.'

— 'Human beings have basic needs: food, shelter, clothing, health, education. Any process of growth that does not lead to their fulfilment - or, even worse, disrupts them - is a travesty of the idea of development.'

—'We are all in need of a redefinition of our goals, or new development strategies, or new lifestyles, including more modest patterns of consumption among the rich.'

The Cocoyoc Declaration ends:

'The road forward does not lie through the despair of doomwatching or through the easy optimism of successive technological fixes. It lies through a careful and dispassionate assessment of the 'outer limits', through cooperative search for ways to achieve the 'inner limits' of fundamental human rights, through the building of social structures to express those rights, and through all the patient work of devising techniques and styles of development which enhance and preserve our planetary inheritance.'

In terms of climate change, growing concern about global warming led to the first World Climate Conference in Geneva in February 1979. It concluded that anthropogenic carbon dioxide emissions could have a long-term effect on climate. The World Climate Programme (WCP) was established the following year, providing the framework for international cooperation in research and the platform for identifying the important climate issues of the 1980s and 1990s, including ozone depletion and global warming.

The defining political events of the 1980s were the breakdown of the Eastern Bloc and the end of the bipolar world built on the balance of power between Western and communist countries and their allies in the developing world. The changes that were the culmination of reform and perestroika in the Soviet Bloc came on the heels of years of apparently strong economic growth and massive military spending.

The situation was markedly different in the developing regions of Africa, West Asia, and Latin America and the Caribbean, where most countries registered little growth in income. The sub-Saharan region fell further behind with per capita income falling 1.2 per cent a year during the 1980s due to a combination of factors, including severe droughts and unfavourable terms of trade.

For many developing countries the 1980s became known as the lost decade. Starting with the debt crisis in Latin America in 1982, the situation was particularly difficult in countries where wars led to the displacement of millions of people. The number of refugees doubled from about 9 million in 1980 to more than 18 million by the early 1990s. Dealing with the cycle of poverty became a particular challenge as population growth in the developing world not only continued but an increasing number of the poor were living in cities. As urban populations grew, cities were finding their physical infrastructure increasingly stressed and unable to cope with demand.

Catching the scientific world as well as policy makers by surprise, measurements by British researchers of the size of the ozone hole were first reported in 1985. The Global 2000 report recognized for the first time that species extinction was threatening biodiversity as an essential component of the Earth's ecosystems. As the interdependence of environment and development became increasingly clear, the United Nations General Assembly adopted the World Charter for Nature, bringing attention to the intrinsic value of species and ecosystems.

Besides new discoveries, the 1980s also saw a range of catastrophic events that left a permanent mark both on the environment and on the understanding of its connection to human health. In 1984, a leak from a Union Carbide plant left 3,000 people dead and 20,000 injured in Bhopal, India. The same year, up to 1 million people starved to death in Ethiopia. In 1986, the world's worst nuclear accident happened as a reactor at the Chernobyl nuclear power plant exploded in the Ukrainian Republic of the Soviet Union. The 1989 spill of 50 million litres of oil from the Exxon Valdez supertanker into Alaska's Prince William Sound demonstrated that no area, however remote and 'pristine', is safe from the impact of human activities.

The World Conservation Strategy

The events referred to above confirmed that environmental issues are systemic and addressing them requires long-term strategies, integrated action and the participation of all countries and all members of society. This was reflected in the World Conservation Strategy (WCS), one of the seminal documents which served to redefine environmentalism post-Stockholm. Launched in 1980 by IUCN, the strategy recognized that addressing environmental problems calls for long-term effort and the integration of environmental and development objectives.

The WCS envisaged governments in different parts of the world undertaking their own national conservation strategies, meeting one of the objectives of Stockholm to incorporate environment in development planning. Since 1980, more than 75 countries have initiated multi-sector strategies at national, provincial, state and local levels. These are aimed at addressing environmental problems such as land degradation, habitat conversion and loss, deforestation, water pollution and poverty.

The World Commission on Environment and Development

However, communicating the message that environment and development were interdependent required a process which carried authority and credibility to the North and South, to government and the business sector, to international organizations and civil society. In 1983, the World Commission on Environment and Development (WCED), also known as the Brundtland Commission, was formed to hold hearings across the globe and produce a formal report of its findings.

The report was issued after three years of hearings with government leaders and the public worldwide on environment and development issues. Public meetings were held in both developed and developing regions, and the process empowered different groups to articulate their views on issues such as agriculture, forestry, water, energy, technology transfer and sustainable development in general. The Commission highlighted environmental problems such as global warming and ozone layer depletion that were new at the time. It expressed concern that the rate of 'change is outstripping the ability of scientific disciplines and our current capabilities to assess and advise'.

The Commission concluded that existing decision-making structures and institutional arrangements, both national and international, simply could not cope with the demands of sustainable development:

'The present decade has been marked by a retreat from social concerns. Scientists bring to our attention urgent but complex problems bearing on our survival: a warming globe, threats to the Earth's ozone layer, deserts consuming agricultural land. We respond by demanding more details, and by assigning the problems to institutions ill equipped to cope with them'.

Thus were the seeds of broader engagement in environment and development issues sown. As a sign of a strengthening non-governmental sector, many new organizations were formed. In Europe, green parties entered the political arena and the membership of grass-roots environmental organizations increased rapidly.

In the wake of the industrial accidents of the 1980s, the pressure on corporations grew. In 1984, UNEP coorganized the World Industry Conference on

Environmental Management, and in 1984 Responsible Care was established by the chemical industry in Canada, one of the first attempts to provide a code of conduct to sound environmental management in the business sector.

By the end of the decade, the concept of eco-efficiency was being introduced into industry as a means of simultaneously reducing environmental impact while increasing profitability. Few if any of these interests were shared by corporations based in developing countries, but there were already debates on the implications of industries migrating to 'pollution havens' in the South. As it became clear that an increasing number of actors would need to grapple with the environmental dimensions of activities previously not recognized as having environmental implications, academic interest in the subject grew.

Environment and development became legitimate subjects of study in more of the established social and natural disciplines, but also new disciplines were born to address cross-cutting areas. Environmental economics, environmental engineering and other previously fringe subjects started to become established and legitimate fields of scholarship, developing their own theories but also proving their worth in real world contexts.

The environment and sustainability still did not feature high in the principles and particularly practice of bilateral aid. As an early sign of change, in 1987 the OECD established a Development Advisory Committee charged with the establishment of guidelines for the integration of environment and development in development assistance programmes. The successful conclusion of the Montreal Protocol in 1987 was held as a promising model for cooperation between North and South, government and business to address global environmental matters. However, dealing with ozone depletion was more straightforward than dealing with other environmental issues that entered the public agenda in the 1980s, notably climate change.

The Intergovernmental Panel on Climate Change (IPCC)

In 1989, the Intergovernmental Panel on Climate Change (IPCC) was established with three working groups focused on the scientific assessment of climate change, environmental and socio-economic impacts, and response strategies, foretelling the wide range of challenges that humanity would face as it entered the last decade of the millennium. The establishment of the IPCC by UNEP and the World Meteorological Organization (WMO) helped develop a broad consensus on the science, social impacts and best responses to human-induced global warming.

The IPCC has contributed immensely to public understanding of the dangers of global warming, particularly in the industrialized countries. In many developing countries, where climate studies are rare and climate experts almost non-existent, climate change is not regarded in the same light. This has led some organizations in the developing regions to complain about 'an enormous disparity in North-South participation. ... Southern countries have no coordinated national climate programmes, few climate researchers, and hardly any data to compute long-term climate projections'.

Multilateral environmental agreements

Some of the major Multilateral Environmental Agreements (MEAs) of the 1980s are:

— the 1982 United Nations Convention on the Law of the Sea (UNCLOS);

— the 1987 Montreal Protocol on Substances that Deplete the Ozone Layer (implementing the 1985 Vienna Convention for the Protection of the Ozone Layer); and

— the 1989 Basel Convention on the Control of Transboundary Movements of Hazardous Wastes and their Disposal (Basel Convention).

Although UNCLOS was signed in 1982, it did not come into force until 12 years later, an indication perhaps of the complexity of negotiating MEAs. The convention, which has 136 parties, is a landmark legal undertaking that encompasses a wide range of maritime issues, including environmental protection. Its environmental provisions include:

— the extension of sovereign rights over marine resources, such as fish, within the 200-mile exclusive economic zones (EEZs);

— obligations to adopt measures to manage and conserve natural resources;

— a duty to cooperate regionally and globally with regard to environmental protection and research related to this protection;

— a duty to minimize marine pollution, including land-based pollution; and

— restrictions on marine dumping by ships.

The Montreal Protocol

The Montreal Protocol to the Vienna Convention on

Substances that Deplete the Ozone Layer, which entered into force in 1989 and had 182 parties as of December 2001, is one of the most successful examples of international environmental cooperation. The success of the protocol in part has hinged upon the Multilateral Fund as an inducement to participation by developing countries. Parties to the Montreal Protocol must annually provide statistical data on the production, import and export of those ozone-depleting substances (ODS) that are controlled by the protocol to the secretariat, through national reports. Reporting rates are high, with more than 85 per cent of the parties reporting their data. The implementation of the protocol has been tightened and expanded significantly over the years, through the 1990 London, 1992 Copenhagen, 1997 Montreal and 1999 Beijing Amendments.

The Basel Convention

The Basel Convention, which entered into force in 1992 and had 149 parties as of December 2001, has three key objectives:

- to reduce transboundary movements of hazardous wastes;
- to minimize the creation of such wastes; and
- to prohibit their shipment to countries lacking the capacity to dispose of hazardous wastes in an environmentally sound manner.

The convention arose out of growing concerns over shipments of waste from industrialized to developing states. Concerned about shipments to Africa, member states of the Organization of African Unity (OAU) responded with the 1991 Bamako Convention on the Ban of the Import into Africa and the Control of the Transboundary Movement and Management of Hazardous Waste within Africa. It came into force in April 1998.

The 1990s were characterized by the search for increased understanding of the concept and significance of sustainable development. This was accompanied by accelerating trends towards globalization, particularly with regard to trade and technology. The conviction grew that there were an increasing number of global environmental problems that required international solutions. The profile of environmental issues was also increasing in the South as new organizations began demanding diagnoses and solutions for developing countries.

The Regional Environmental Centre was established in Hungary in 1990 to address environmental issues in post-Soviet Central Europe. There was significant action by private industry to put its house into better environmental shape and explosive growth in the use of the Internet and electronic communications. The decade started badly for the environment with the loss of thousands of lives in the 1991 Persian Gulf conflict and a partial black-out over some of the area as millions of barrels of oil were wilfully ignited.

For West Asia, this was a major environmental catastrophe. An oil slick - caused by the release of between 0.5 million to 11 million barrels of crude oil - is reported to have killed 15,000-30,000 sea birds. In addition, about 20 per cent of mangroves in the Persian Gulf were contaminated and 50 per cent of coral reefs affected. The atmosphere was not spared either: about 67 million tonnes of oil were burnt, producing about 2.1 million tonnes of soot and 2 million tonnes of sulphur dioxide.

Elsewhere, while technical progress was transforming industrialized society, few in the developing world were benefiting. The death toll from infectious diseases was 160 times greater than the number killed in 1999 from natural disasters, including earthquakes in Turkey, floods in Venezuela and cyclones in India. The International Federation of Red Cross and Red Crescent Societies reported that a 1995 survey of 53 countries showed a 15 per cent decline in health spending per person following economic structural adjustment.

By 1997, near the end of the 20th century, some 800 million people not only went hungry every day but also lacked the basic skills of reading and writing essential to sustainable development. In terms of governance, events of the late 1980s continued to influence political developments across the globe. No region was immune as dictatorships and military regimes in Africa and Latin America were voted out of power, and the single party governments in some European countries were relegated to opposition benches by a restless electorate.

The people had begun to exercise their right to elect their leaders and demand accountability. Despite this radical change in terms of governance, there was little immediate impact on the environment in most countries. In the countries of the former Soviet Union, however, economic recession helped reduce waste emissions and energy consumption. Whether such effects will prove only temporary remains to be seen.

At the institutional level, ideas that had taken shape during the late 1980s, such as multistakeholder participation and increased accountability on environmental and social matters, were given a higher profile by a number of international events. The first of these was a ministerial conference on the environment held in Bergen, Norway, in May 1990, where such ideas were first formally endorsed. This conference was convened to prepare for the UN Conference on Environment and Development (UNCED or the Earth Summit) that was held in June 1992 in Rio de Janeiro, Brazil.

The Earth Summit

UNCED was attended by an unprecedented number of representatives of state, civil and economic society -176 governments, more than 100 heads of state compared to the two who attended the 1972 Stockholm Conference, and an estimated 10,000 delegates, 1 400 nongovernmental organizations (NGOs) and about 9,000 journalists. It is still the largest such gathering ever held. Prior to the Summit itself, the preparatory events from national and sub-regional to regional and global also involved the participation of hundreds of thousands of people across the world, ensuring that their voices were heard. Sub-regional and regional organizations such the Association of Southeast Asian Nations (ASEAN), the Organization of African Unity, the European Union and many others played a role before and during the Earth Summit. They continue to do so in trying to implement Agenda 21, the action programme that resulted from the conference. The Summit produced at least seven major achievements:

— the Rio Declaration on Environment and Development (containing 27 principles);

— Agenda 21 - a blueprint for environment and development into the 21st century;

— two major international conventions - the United Nations Framework Convention on Climate Change (UNFCCC) and the Convention on Biological Diversity (CBD);

— the Commission on Sustainable Development (CSD);

— agreement to negotiate a world desertification convention; and

— the statement of Principles for the Sustainable Management of Forests.

The Rio Principles reaffirmed the issues that had been articulated in Stockholm 20 years earlier, placing human beings at the centre of sustainable development concerns by stating that humans are 'entitled to a healthy and productive life in harmony with nature'. The Earth Summit provided a forum to address issues of both environment and development, and to highlight differences in perspective between the North and South.

After the Summit, sustainable development took on a life of its own, forcing its way into the deliberations of bodies ranging from city councils to international organizations. More than 150 countries have established national-level institutions to develop an integrated approach to sustainable development -although, in some countries, the national councils of sustainable development have been more political than substantive in nature.

A wide range of civil society sectors is now involved in agenda setting and strategy building. More than 90 per cent of them have been established in response to Rio, the majority in developing countries. The emphasis given to sustainable development also had considerable impact on both legal instruments and the institutions that administer them. For example CITES, which was already evolving away from a classic conservation approach, moved further towards an approach that balances conservation with sustainable use. The practical application of sustainable use within CITES provoked substantial and heated debate throughout the decade.

Agenda 21

Agenda 21 is an action programme. It was partially based on a series of specialized contributions from governments and international bodies, including Caring for the Earth: a Strategy for Sustainable Living. Agenda 21 is now the most significant and influential non-binding instrument in the environmental field, serving as the blueprint for environmental management in most regions of the world. The cost of implementing Agenda 21 in developing countries was estimated by the Earth Summit secretariat to be about US$625 billion a year, with the developing countries meeting 80 per cent, or US$500 billion, of it.

Developed countries were expected to foot the remaining 20 per cent, or about US$125 billion annually, by meeting their long-established official development assistance (ODA) target of 0.7 per cent of gross national product (GNP). Although UNCED was concerned with global approaches, an important outcome was the adoption of many national and regional Agenda 21 programmes for sustainable development. In the Southern African Development Community (SADC) region, for example, the member states adopted the SADC Policy and Strategy for Environment and Sustainable Development in 1996. The European Union adopted the 5th Environmental Action Plan Towards Sustainability.

The Global Environment Facility

The Global Environment Facility (GEF) was created in 1991 as an experimental partnership involving UNEP, UNDP and the World Bank to generate ecological dividends from local and regional development by providing grants and low-interest loans to developing nations and economies in transition. Following the Summit, it was intended to be the financing mechanism for Agenda 21, and was expected to mobilize the necessary resources. It helps fund national, regional and global development projects which benefit the world's environment in four focal areas - climate change, biodiversity, ozone and international waters - as well as local economies and societies.

UN Framework Convention on Climate Change

The ability of IPCC to provide evidence that climate change posed a real threat encouraged governments at the Summit to sign the UN Framework Convention on Climate Change (UNFCCC). This became the centrepiece of the Summit and entered into force in 1994; it had 186 parties as of December 2001. The development of the convention can be traced back to the 1990 Second World Climate Conference in which the ministerial declaration was the vehicle for pushing ahead with policy development and the establishment of the Global Climate Observing System (GCOS).

The primary goal of the UNFCCC is to stabilize greenhouse gas emissions at levels that will prevent dangerous anthropogenic interference with the global climate. The principle of 'common but differentiated responsibilities' that was embraced by this convention has guided the adoption of a regulatory structure. This principle reflected the reality that most emissions of greenhouse gases are from industrialized states. The Kyoto Protocol, which set actual targets for emissions reductions, was opened for signature in 1997. As of December 2001, 84 Parties had signed and 46 Parties had ratified or acceded to the Protocol. A notable exception was the United States which announced its decision not to ratify the Protocol in early 2001.

The Convention on Biological Diversity

The CBD came into force in 1993. It was the first global agreement on the conservation and sustainable use of biodiversity and serves as a blueprint for national action. The Convention establishes three main goals: the conservation of biological diversity, the sustainable use of its components, and the fair and equitable sharing of the benefits from the use of genetic resources. Many biodiversity issues are addressed, including habitat preservation, intellectual property rights, biosafety and indigenous peoples' rights.

The Convention stands as a landmark in international law, noted for its comprehensive, ecosystems approach to biodiversity protection. The treaty has gained rapid and widespread acceptance. By December 2001, a total of 182 governments had ratified the agreement. A supplementary agreement to the Convention, the Cartagena Protocol on Biosafety, was adopted in January 2000 to address the potential risks posed by cross-border trade and accidental releases of living genetically modified organisms.

The adoption of the biosafety protocol is a success for developing countries which called for it. The protocol had been signed by 103 parties and ratified by 9 as of

December 2001. The CBD has also influenced the enactment of a law which seeks to regulate genetic resources within the Andean Pact nations of Bolivia, Colombia, Ecuador, Peru and Venezuela. The law became effective in July 1996. Despite the success of the convention, negotiations before its adoption were often acrimonious.

The Convention to Combat Desertification

Though negotiations were not completed until 1994, the UN Convention to Combat Desertification (CCD) developed out of the process associated with the 1992 UNCED. However, its history extends back to the 1970s. It became effective in 1996 and had 177 parties as of December 2001. The CCD has been described as a 'Rio stepchild' because it did not get as much attention as the UNFCCC and CBD. The industrialized countries opposed it because 'they were not willing to undertake any financial responsibility for arresting the process of desertification' which is not perceived as a global problem.

While it has been projected that a 20-year global effort to combat desertification would cost US$10-22 billion annually, funding countries provided a meagre US$1 billion in 1991 for desertification control worldwide. Although the CCD has modest financial support compared to the UNFCCC and the CBD, the convention is distinctive for two reasons:

— It endorses and employs a 'bottom-up' approach to international environmental cooperation. Under the terms of the CCD, activities related to the control and alleviation of desertification and its effects are to be closely linked to the needs and participation of local land users and nongovernmental organizations.

— It employs detailed regional annexes, sometimes more detailed than the core treaty itself, that address the particularities of the desertification problem in specific regions such as Africa, Latin America and the Caribbean, and the Northern Mediterranean.

The central substantive commitment in the CCD is the obligation to develop 'national action programmes', in conjunction with local stakeholders. These programmes delineate the tasks that the parties will undertake to implement the CCD. For example, parties must make the prevention of desertification a priority in national policies and must promote awareness of desertification among their citizens.

The establishment of the Commission on Sustainable Development (CSD) in December 1992 was a direct outcome of the Summit. Although the goal of sustainable development was established in the 1980s, it was not until Rio that an international body was assigned the mandate to oversee and help the international community achieve this goal. Although this was a major step, the Commission has been criticized for being a weak response to problems of institutional capacity, and has encountered apathy from high-level government ministers. The integration of economic, social and environmental policies - a requirement of sustainable development spelled out by the Brundtland Commission -continues to challenge institutions at all levels.

The principles of sustainable development were reaffirmed throughout the 1990s at numerous international conferences such as the:

— 1993 World Conference on Human Rights in Vienna;

— 1994 International Conference on Population and Development in Cairo;

- 1994 Global Conference on the Sustainable Development of Small Island Developing States in Bridgetown, Barbados;
- 1995 World Summit for Social Development in Copenhagen;
- 1995 Fourth World Conference on Women in Beijing;
- 1996 United Nations Conference on Human Settlements (Habitat II) in Istanbul; and
- 1996 World Food Summit in Rome.

Stakeholder participation in sustainable development

Much of this international activity was mirrored by attempts by the private sector to improve its environmental performance. Action was encouraged by the creation of the World Business Council for Sustainable Development (WBCSD) in 1995 which has done much to encourage industry to look for improvements in profitability by reducing waste of both resources and energy and by reducing emissions. The WBCSD now has hundreds of members, of whom many have managed to effect remarkable savings for both themselves and the environment.

In 1996, the International Organization for Standardization created a new voluntary standard for environmental management systems in industry, the ISO 14,000. By the end of the decade, transnational corporations had greatly improved their environmental image; indeed, their environmental performance was often better than that of many small and medium-sized enterprises. Corporate environmental reporting also became more common during the 1990s and the Global Reporting Initiative was created to establish a common framework for voluntary reporting of the environmental, economic and social performance of an organization.

The GRI seeks to elevate enterprise-level sustainable development reporting to the same level of credibility, comparability and consistency as financial reporting. Civil society was also active, notably in its attempts to create an Earth Charter which articulates the 'fundamental ethical principles for a sustainable way of life'. Hundreds of groups and thousands of individuals have been involved. The Charter, which was originally meant to have been adopted at the Earth Summit, has been refined in a process spearheaded by the Earth Council and Green Cross International.

Civil society has not, however, limited itself to campaigns such as the Earth Charter but has also organized massive demonstrations in different parts of the world, many of them against the perceived threat of globalization. Such attempts are themselves reflections of the globalization process, and of the now extraordinary power of the Internet which has undergone explosive growth. While in 1993, there were only 50 pages on the World Wide Web, these had multiplied by a million by the end of the decade, making radical changes to the way many people live and work - mainly in the rich industrialized countries.

Even though 'electrons are cheap', at the end of the 1990s 88 per cent of Internet users lived in industrialized countries, which collectively represented just 17 per cent of the world's population. This was a sobering conclusion to the end of the 1990s: in at least one important sense, the voices and concerns of the poor majority - for all the decade's rhetoric - were still being left out of the global conversation.

The Comprehensive Nuclear Test Ban Treaty

An important milestone in international cooperation with a bearing on the environment came in 1996 with the adoption of the Comprehensive Nuclear Test Ban Treaty (CTBT) by the United Nations General Assembly in New York. The CTBT, which prohibits all nuclear test explosions in all environments, was opened for signature in New York on 24 September 1996, when it was signed by 71 States, including the five nuclear weapon states. As of August 2001, 161 States had signed the Treaty and 79 had ratified it.

Thirty years ago, many toxic and persistent chemicals were considered to be not only resources but also pollutants adversely affecting human health and the environment, particularly where they could be accumulated up the food chain or transported long distances over the globe. Today, chemicals are seen as even more essential to development, and as a resource that needs to be managed to protect or even enhance human health and the environment. This sound management of chemicals applies to both those anthropogenically produced and those of natural origin, including those generated through biological processes.

The international community has recently concluded a landmark convention to control the use of a group of persistent toxic organic compounds. In December, 2000 representatives of 122 governments finalized the text of a legally binding treaty for implementing international action on certain persistent organic pollutants (POPs). The

Stockholm Convention on Persistent Organic Pollutants, signed in May 2001 and which had 111 signatories and 2 Parties as of December 2001, sets out control measures covering 12 chemicals. The control provisions call for eliminating production and use of intentionally produced POPs and eliminating unintentionally produced POPs where this is feasible.

Since the Stockholm Conference, the global chemicals industry has grown almost ninefold and an annual growth rate of about 3 per cent is expected to continue over the next three decades, with a considerable increase in trade. This will increase the risk of exposing an increasing number of people and the environment to new chemicals and the potential for the emergence of new diseases of chemical origin. Information about the release of chemicals into the environment is now much more widely available than used to be the case.

North America has led action in this area, in particular with the US Toxics Release Inventory enacted through the Emergency Planning and Community Right-to-Know Act (EPCRA) in the United States in 1986. EPCRA's purpose is to inform communities and citizens of chemical hazards in their areas. The Act requires businesses to report the locations and quantities of chemicals stored on-site to state and local governments. Through EPCRA, the US Congress mandated that a Toxics Release Inventory (TRI) be made public.

TRI provides citizens with information about potentially hazardous chemicals and their use so that communities have more power to hold companies accountable and make informed decisions about how toxic chemicals are to be managed.

Environmental issues featured prominently during the United Nations Millennium Summit hosted by Secretary-General Kofi Annan in New York in 2000. While recognition of the importance of environmental issues at this summit was encouraging, the actual progress report was not. The Secretary-General was blunt in his comments regarding environmental management, stating that the international community was failing to provide future generations the freedom to 'sustain their lives on this planet. On the contrary', he said, 'we have been plundering our children's future heritage to pay for environmentally unsustainable practices in the present'.

In early 2001, IPCC announced that the evidence for anthropogenic climate change was getting stronger, that warming was happening faster, and that the consequences looked more severe than first predicted. The expert panel, made up of thousands of scientists from around the world, predicted that average temperatures would rise between 1.4 and 5.8°C over the 21st century.

IPCC states that: 'There is new and stronger evidence that most of the warming observed over the last 50 years is attributable to human activities ... Furthermore, it is very likely that the 20th century warming has contributed significantly to the observed sea level rise, through thermal expansion of sea water and widespread loss of land ice'. The implications of this faster rise in global temperatures cut across broad areas of economic, social and environmental concern and add urgency to the need for controls on factors contributing to global warming.

Globalization has been described by some as the new Industrial Revolution. There are fears that it could result in a dangerous polarization between people and countries benefiting from the system and those that are merely passive recipients of its effects. In its Human Development Report 1999, UNDP states that the one-fifth of the world's people living in the highest income countries have 86 per cent of world GDP, 82 per cent of world export markets, 68 per cent of foreign direct investment, and 74 per cent of telephone lines. The bottom one-fifth, in the poorest countries, have about 1 per cent in each category. In the 1990s, more than 80 per cent of foreign direct investment in developing countries and those of Central and Eastern Europe went to just 20 countries, notably China.

The main engines of globalization are the transnational corporations, transnational media organizations, intergovernmental organizations and NGOs. Globalization is more than the flow of money and commodities - it is the growing interdependence of the world's people through 'shrinking space, shrinking time and disappearing borders'. This offers great opportunities for enriching people's lives and creating a global community based on shared values.

One response to the rapid growth of economic globalization has been massive civil society demonstrations in many parts of the world. The November 1999 demonstration at the World Trade Organization (WTO) meeting in Seattle in which thousands of people campaigned against globalization was a major event. Increased public awareness was also apparent at the World Bank and International Monetary Fund meetings in Prague in September 2000, and at other meetings since. These demonstrations show that citizens across the globe are insisting on being heard, and are demanding the implementation and enforcement of acceptable trade, labour and environmental standards worldwide.

Many international organizations involved in regulating the global economy are now having to adjust their policies to include the participation of civil society in their activities. The irony of globalization and increased public awareness is that consumption by people in industrialized countries continues to grow, and poverty in developing regions continues to worsen.

Towards Sustainable Development

The year 2002 is the beginning of the fourth decade since the international community laid the foundation in 1972 for collective global action to mitigate adverse impacts on the environment. It finds one of the three pillars of sustainable development—the environment-seriously listing because of the distortions placed on it by the actions of a human population that now numbers more than 6 000 million. The importance of the environment is often underplayed even though its value to human survival and development is incalculable. The collapse of the environmental pillar is a serious possibility if action—from local to global—is not taken as a matter of urgency to address human impacts, which have left:

— increased pollutants in the atmosphere;

— vast areas of land resources degraded;

— depleted and degraded forests;

— biodiversity under threat;

— increasingly inadequate freshwater resources of deteriorating quality; and

— seriously depleted marine resources.

The environment is under siege. Unless both shortand long-term changes are instigated, sustainable development will remain a chimera—possibly only in the haze on a distant horizon. There is need for a balanced approach towards sustainable development. All three pillars - social, economic and environmental—are mutually supportive and all three are essential. Neglecting any one, and this is all too frequently the case with the environmental pillar, is not only shortsighted but leads to a policy dead end. The disintegration of the environmental pillar will lead to the inevitable collapse of the other, more charismatic pillars of sustainable development to which policy makers everywhere pay particular attention.

The world is now split into the haves and the havenots by four major divisions, all of which continue to widen. They are:

— the environmental divide;

— the policy divide;

— the vulnerability gap; and

— the lifestyle divide.

These four divisions are a serious threat to sustainable development. Over the past three decades, massive investments of human and financial resources have been used to exploit the environment. On the other hand, research has opened up new frontiers in terms of humanity's understanding of the complex web of ecological processes. Policies have been introduced to address many of the key issues. Targets have been set and met in some areas, such as the phase out of ozonedepleting substances, but success has been limited in others, for example the adoption of more stringent targets to reduce anthropogenic greenhouse gas emissions under the Kyoto

Protocol. Many other initiatives critical to closing the lifestyle divide and the success of sustainable development have been identified. These include:

Alleviating poverty

The international community has set a target of halving by 2015 the proportion of the world population which survives on less than US$1 a day. The day-to-day lives of the majority of the poor are much more closely linked to the state of the environment than is the case for the better off -a healthy, productive environment is one of the few stepping stones out of poverty. As long as millions of the world's population remain poor, and the environment stays on the periphery of mainstream policy making, sustainable development will be an unachievable ideal. Reducing the excessive consumption of the more affluent.

As long as the richest 20 per cent of the world population continues to account for 86 per cent of total personal consumption expenditure, it is unlikely that sustainable development will ever be achieved. The resulting pockets of wealth in a sea of poverty heighten tensions and overexploit resources. Improving governance has become a major issue, not only at the institutional and national levels, but also at the global level where disparities between North and South often fuel conflict and intransigence in negotiating policies for effective environmental management.

Eliminating debt, particularly for the Highly Indebted Poor Countries, is also an important factor in a world in which debtor nations often use more foreign currency to repay debts than they earn. Indebtedness frequently leads to the overexploitation of the environment. As long as this situation is perpetuated, many of the debtor nations are unlikely to ever achieve sustainable development. Many complex processes - social, economic and environmental - are at play in terms of greater impacts on people as a result of environmental change.

The main difference is in coping capacity and ability to recover, and this difference means that the poor are generally more vulnerable. This vulnerability gap is undermining sustainable development. Human vulnerability to environmental change encompasses the quality of the environment, threats to that environment and the differing coping strategies of individuals and communities in any location, country or region. Closing the vulnerability gap would have a huge impact on the well-being and security of millions. Reducing and eliminating poverty are inextricably linked with sound environmental management, which includes such issues as property and usage rights, the provision of basic services to protect the environmental asset base, adequate infrastructure, and funding for development and environment activities.

One way forward here would be for donors to give direct support to community-based initiatives, especially those channelled towards sustainable development activities, through funding channels accessible to low-income, disadvantaged and vulnerable groups. Changes in consumption levels are needed by the more affluent individuals and nations of the world. Prosperity is closely linked to the ability to address environmental problems but it is also one of the forces behind excessive consumption, which is the cause of other problems with far-reaching impacts.

Economic and political concerns have stalled attempts to change consumption patterns through new policies or instruments. A realization that changing consumption patterns does not have to curtail or prejudice quality of life, and can in fact do the opposite, must be brought home to the people concerned. There is sufficient evidence that this is the case but no coordinated effort to get the message across has yet been undertaken.

Changing mindsets needs to go hand in hand with increased acceptance of responsibility for environmental and social impacts, and the creation of consumer ethics. The provision of financial resources is insufficient in the absence of adequate capacity. Targeted capacity building and, more importantly, capacity mobilization and retention to minimize the brain drain are needed for more effective environmental governance and public participation. It is particularly important to ensure that capacity development is a shared experience rather than top-down or North-to-South instruction.

With enhanced capacity, developing regions may be better able to cope with environmental change and disasters, which have increased their vulnerability. The following are some of the additional environment-related challenges policy makers at all levels face over the coming decades:

— Large numbers of people, especially in developing countries, in both rural and urban areas, still lack access to clean water and adequate sanitation, good outdoor and indoor air quality, cleaner energy and waste management. This continues to lead to the degradation of the natural assets base, ill health and vulnerability to environmental threats.

— Unresolved conflicts remain over the ownership and management of common property resources (such as water, air, land, forests and oceans).

— Highly complex environmental issues which are not yet adequately addressed include the increasing prevalence of persistent toxic substances, unsafe handling, disposal and dispersal of chemical and hazardous wastes, non-point pollution sources, management of transboundary river systems and shared water bodies, and excessive nitrogen loading.

— Climate change will cause inevitable damage in the medium and long term (low-lying islands and coastal areas, arid and semi-arid ecosystems, increased scale and intensity of environmental disasters). Developing countries, particularly Small Island Developing states, are the least able to adapt to events caused by climate change but are the most likely to be affected by them.

— The global environmental impact (ecological footprint) of the developed world and prosperous communities elsewhere is larger than that of the poor in the developing world but future economic development and population growth in the latter are likely to dramatically increase environmental impacts.

Impact of information

Information is the foundation of sustainable development and is fundamental to successful planning and decision making. If decisions are made without sound data and information, they will be little better than best guesses and are likely to be wrong. Economic and social data are widely available and are relatively reliable and well understood. The situation with environmental data and information is somewhat different. High quality, comprehensive and timely information on the environment remains a scarce resource, and finding the 'right' information can pose problems: data are more difficult and expensive to obtain.

It is also difficult to find indicators that capture and reflect the complexity of the environment and human vulnerability to environmental change. Environmental data acquisition remains a basic need in all countries. Despite the problems, scientific consensus based on best available data and knowledge, although sometimes partial and limited, has proved a powerful tool for bringing environmental issues to international attention, and prompting action. There have been great improvements in environmental research and monitoring but it is important that these efforts are maintained and improved to ensure a flow of timely and reliable information. A sub-set of that information needs to underpin early warning in relation to disasters and emerging issues, and to the factors underlying human vulnerability to environmental change. The provision of information will benefit from:

— Structured and nested monitoring and observing systems including global satellites for collection of timely and reliable data on environmental components, using harmonized units of measurement and terms.

— An effective, globally acceptable environmental information system supported by a harmonized set of data, indicators and indices and closely integrated with socio-economic information systems to provide basic information for decision making.

— Maximizing the use of the Internet as a cheap and effective means of information exchange worldwide. Identifying indicators and indices to capture the sustainability of environmental trends as well as trends in human vulnerability to environmental change.

— Tools to create a knowledge base accessible to environmental policy and decision makers. Presentation of complex data and information in an easily understandable form to decision makers.

Many environmental institutions were originally set up under different conditions and to perform different functions from those they are expected to exercise today. They now need to adapt to new roles and partnerships, especially in view of the development of multiple, nested levels of action from the global to the local, and the spread of responsibilities to civil society. Many institutions are constrained by a lack of human capacity and funding, despite increased environmental challenges, and this limits their effectiveness. These are clearly issues that need to be addressed if institutions are to fulfil their present obligations and confront emerging environmental issues.

The most successful environmental management approaches are based on policies that incorporate a range of tried and tested principles. Policy development processes for the environment need tightening to become more rigorous, systematic and integrated, turning out policies tailored for specific localities and situations. This implies a clearer understanding of the environmental implications of decision-making processes, better information on the quality of environmental management, the prioritization of issues, the setting of goals, targets and time frames, and the development of tools for monitoring progress and evaluating results.

Environmental policy development is a dynamic, iterative process encompassing assessment of the problems and options, target setting and policy formulation, followed by implementation, monitoring, review, evaluation, regular reassessment and adjustment. Without the complete cycle, it is not possible to determine if policies are having their intended effect or to make informed adjustments. Effective environmental policies form part of mainstream politics and should be incorporated into all political sectors and levels of government, especially into economic decision making.

Effective national plans and sustainability strategies bring together governments, civil society and the private sector; they incorporate diverse initiatives such as 'green' and 'brown' agendas and country-specific best practices. Achieving this implies shifting the focus of policy from environmental protection to the broader concept of sustainable resource management.

The fragmentation, duplication and overlap among international Multilateral Environmental Agreements (MEAs) reflects their piecemeal origin in response to each new problem. There are more meetings and reporting requirements than many governments can cope with or afford. Compliance rates are at best patchy. In the long term, a different approach aimed at building a coherent body of international environmental legislation is needed; meanwhile, short-term initiatives may help the system function more effectively.

Benefit of Sustainable Development

Western trade barriers restrict the ability of developing countries to grow economically and finance sustainable development expenditure. Global trade patterns and increasing globalization can threaten the environment but they can also provide mechanisms and incentives for sustainable development. The open market brings new opportunities for flows of capital, technology and labour to benefit developing countries. The danger of trade liberalization, however, is that environmental policies can be undermined and pollution havens created unless adequate checks and balances are in place.

Some form of sustainable trade is therefore needed. Sustainable trade is defined as an international exchange of goods and services that yields positive social, economic and environmental benefits, and reflects the core criteria of sustainable development. If trade, environment and development are to be brought together successfully, three critical needs must be addressed:

— the need for transparency;

— the need for equity and fairness - especially so that countries or stakeholders historically excluded from trade opportunities really benefit; and

— the need to ensure that social and environmental benefits provide lasting value for developing countries in terms of the capacity for innovation in the future.

Promising new technologies can help to secure basic goods and services, especially for the poor in developing countries, but they remain largely the province of commercial interests in developed countries. Environmental technologies in water and energy, recycling, biotechnology and ecological farming all offer great potential. Ways must be found to capitalize on these innovations, transfer the technologies to the needy at affordable cost and include them in the development of technologies.

Developing countries must become 'partners in the process' of technology rather than 'partners in the products' of technology. In spite of the enormous potential for environmental and social gains from new technology, there are certain risks that need to be managed. Methodologies and capacity to evaluate these risks, to establish 'rules of conduct' and to facilitate appropriate transfer provide the challenge. The precautionary approach, polluter and user pays principles, full information dissemination and disclosure of risks, technology impact assessment, and the cost benefit analyses of adopting the technology must all be taken into account.

When environmental goods and services are costed, rather than considered as free, there is an appreciation of the immense value of the environmental assets that underpin human well-being and security. This value is proportionally greater for the livelihoods, survival ability and resilience of the poorest and most marginalized communities. Valuing the environment is one way of raising the profile of the environment in the national policy arena. There have been numerous attempts to develop methods to assign a monetary value to environmental goods and services and the environmental impacts of economic activity.

A plethora of instruments exists that can make the market work for sustainable development, including tradable permit schemes, removing market barriers and environmentally damaging government subsidies, subsidizing the start-up of environmentally sound businesses, creating markets for environmental services, encouraging disclosure policies and recycling tax revenues.

In the right context, market instruments can often be more effective than command-and-control measures.

Furthermore, their flexibility encourages private sector innovation in ways that binding policies do not. The market is not very effective, however, in dealing with the long time horizons and uncertainty that characterize some environmental problems. Voluntary initiatives are increasingly recognized by the private sector as a way of contributing to the environmental agenda and being seen to be 'green'. At the individual level, voluntary action is the foundation for many forms of environmentally responsible behaviour.

Governments are often good at adopting declarations and action plans but not at implementing them. The lack of political will, implementation, enforcement and compliance requires urgent action. Open processes of performance monitoring are often absent and the responsibilities for it at each level of governance unclear. NGOs play an essential role in independent review at national level, drawing attention to policy failures and poor performance.

The creative interaction of individuals and small units often provides efficient solutions to managing complicated and variable situations. Local responsiveness and adaptability are important. This approach can be applied at a variety of scales and has important implications for environmental management, suggesting the need to distribute roles and responsibilities in new ways. The transfer of certain responsibilities through subsidiarity and decentralization is emerging as an effective way to ensure more timely policy development and implementation.

Responsibility for many aspects of environmental and social health and safety lies at the local or municipal level, where action is crucial for poverty reduction, improving local environments and providing early warning on issues with current or potential national and international prominence. The effectiveness of devolving power to this level depends on the nature of participatory management systems of environmental governance, identifying all stakeholders and ensuring that they are 'at the policy table'.

Particularly in developing countries, providing for more meaningful participation in environment and resource use decision-making, and giving all stakeholders the confidence that they can make a difference, will decrease mutual suspicion and enable major groups to participate in managing the shared environment on an equal footing.

Regional levels of environmental policy development are also important, requiring the strengthening of regional bodies, initiatives, institutes and alliances. Strengthening regional level roles creates capacity for successful policy at an appropriate scale for many environmental processes and threats. In the interlinked chain from local to global levels, regional blocks are becoming more important, new ones are being formed and stalled initiatives are being given new life. Increased attention is needed to develop regional level environmental policy, with a clear definition of responsibilities among local, national, regional and global institutions.

There is a growing range of global stakeholders including intergovernmental organizations, multinational corporations and international NGOs. Environmental NGOs are increasingly involved in international environmental governance, including the implementation of activities under the Global Environment Facility, the World Bank and regional development banks.

5

Coastal and Marine Environment

Marine and coastal degradation is caused by increasing pressure on both terrestrial and marine natural resources, and on the use of the oceans to deposit wastes. Population growth and increasing urbanization, industrialization and tourism in coastal areas are root causes of this increased pressure. In 1994, an estimated 37 per cent of the global population lived within 60 km of the coast—more people than inhabited the planet in 1950. The effects of population are multiplied by both poverty and human consumption patterns.

Prior to 1972, the crash of some seabird populations caused by DDT, outbreaks of Minamata disease in Japan from mercury-contaminated seafood, and the Torrey Canyon and other oil spills focused the attention of the Stockholm Conference on marine pollution. Policy responses included bans on production and use of some substances, regulations to reduce discharges, and the prohibition of ocean dumping, as well as a significant scientific effort to improve the status of knowledge about these pollutants.

These responses are enshrined in a number of international agreements, including the 1972 London Dumping Convention and its 1996 Protocol, the 1989 Basel Convention on the Control of Transboundary Movement of Hazardous Wastes and their Disposal, and the 1995 Global Programme of Action for the Protection of the Marine Environment from Land-based Activities.

Marine pollution is also an important focus of UNEP's Regional Seas Programmes that have been established in many parts of the world. Globally, sewage remains the largest source of contamination, by volume, of the marine and coastal environment, and coastal sewage discharges have increased dramatically in the past three decades. In addition, because of the high demand for water in urban neighbourhoods, water supply tends to outstrip the provision of sewerage, increasing the volume of wastewater.

Public health problems from the contamination of coastal waters with sewage-borne pathogens were well known in the 1970s, and in many developed countries improved sewage treatment and reduction of the disposal of industrial and some domestic contaminants into municipal systems have significantly improved water quality. In the developing world, however, the provision of basic sanitation, as well as urban sewer systems and sewage treatment, has not kept pace. High capital costs, the explosive pace of urbanization, and in many cases limited technical, administrative and financial capacities for urban planning and management and ongoing operation of sewage treatment systems are barriers to efficient sewage treatment.

Removal of these barriers, as well as alternative approaches, is urgently needed. Recent evidence suggests that bathing in waters well within current microbiological standards still poses significant risk of gastrointestinal disease, and that sewage contamination of marine waters is a health problem of global proportions. A primary concern at the Stockholm Conference was the introduction of nutrients to coastal and marine waters.

Human activities now account for more than half of global nitrogen fixation, and the supply of fixed nitrogen to the oceans has greatly increased. Sewage discharges are often the dominant local source near urban areas but global inputs are dominated by agricultural run-off and atmospheric deposition. At the time of Stockholm, agricultural nutrient run-off was 'not yet a major global problem'. Most fertilizer use was in developed countries but the rapid increase of fertilizer use in developing countries was already foreseen.

Fertilizer use has stabilized in developed countries but is increasing in developing ones, a trend expected to continue. Fertilizer use has undoubtedly been enhanced by widespread subsidies, which reflect the high political priority of increasing food production and reducing food costs. Atmospheric inputs, derived primarily from vehicle and industrial emissions and in some areas evaporation from animal manure and fertilizer, dominate anthropogenic nitrogen inputs to some coastal areas. They are expected to rise with increasing industrialization and vehicle use, especially in developing regions.

Atmospheric nitrogen inputs to the nitrogen-limited open oceans will also increase, with potential significant impacts on primary production and the carbon cycle. Marine and coastal eutrophication from elevated nitrogen inputs has emerged as a worrying trend not foreseen three decades ago. There is increasing evidence that blooms of toxic or otherwise undesirable phytoplankton are increasing in frequency, intensity and geographic distribution.

Severe eutrophication has occurred in several enclosed or semi-enclosed seas, including the Black Sea. Elsewhere, elevated growth and subsequent decay of phytoplankton has caused widespread areas of seasonally oxygendepleted water. Phytoplankton blooms can have major economic impacts on fisheries, aquaculture and tourism. At the time of the Stockholm Conference concerns for ocean health centred on pollution by POPs, heavy metals and oil. Some response measures have been effective, for example, introduction of unleaded gasoline helped to reduce lead levels in Bermuda; national regulations and international agreements such as the Convention on the Prevention of Pollution from Ships (MARPOL) resulted in the reduction of operational oil discharges from ships; and North American seabird populations affected by DDT recovered after this chemical was banned in the region.

In other cases, improved information has clarified some concerns. High levels of mercury in tuna and swordfish, for example, have been shown to have natural sources; the most dramatic effects of oil spills have proved to be localized and relatively transient; and heavy metal contamination, except for lead and mercury, has been found to be highly localized and has relatively minor impacts except at high concentrations.There are, however, other continuing concerns about these pollutants.

The chemical residues of oil spills may have subtle long-term effects, and chronic, small releases cause seabird mortality and other environmental effects. The effects of heavy metal contamination can be severe and are a significant concern in the Arctic. The most serious concerns globally relate to POPs, many of which are transported globally via the atmosphere and are ubiquitous in the oceans.

There is growing evidence that long-term, low-level exposures to some POPs cause reproductive, immunological, neurological and other problems in marine organisms, and possibly in humans, but the evidence for widespread ecological or human health impacts at current levels of contamination remains equivocal. Another threat to the oceans, and in particular to living organisms, is non-biodegradable litter which enters the sea.

Each year, large numbers of seabirds, sea turtles and marine mammals are killed by entanglement in or ingestion of non-biodegradable litter. Human-induced changes in the natural flow of sediment have emerged since the Stockholm Conference as a major threat to coastal habitats. Urban and industrial development drives the construction of residential and industrial infrastructure which, depending on its nature, can alter sediment flow.

In addition, agriculture, deforestation and construction typically mobilize sediments. Deltas, mangrove forests, beaches and other coastal habitats are sustained by the supply of sediment, while other habitats, such as coral reefs and seagrass beds, may be smothered or deprived of light. Sedimentation is one of the major global threats to reefs, particularly in the Caribbean, Indian Ocean, and South and Southeast Asia.

The Stockholm Conference projected that annual harvests could approximately double from 1970 levels to 'rather more than 100 million tonnes', although the depletion of some fisheries by overexploitation was also recognized. In the same year the world's largest fishery, the Peruvian anchovy, crashed spectacularly, a result of years of unsustainable harvests precipitated by a strong El Niño event. Harvests from marine capture fisheries did rise but failed to reach 100 million tonnes, fluctuating around 80-90 million tonnes from the mid-1980s.

Aquaculture production, by contrast, has risen sharply but is entirely dominated by Asia and the Pacific. The Stockholm Conference recommended two basic approaches to fisheries management: improving management information through research, assessment and monitoring, and international cooperation. Despite great improvement in the quality and scope of fisheries information, better fisheries management has generally not been achieved. There has been an almost inexorable global trend towards increasingly intense exploitation and depletion of fisheries stocks, three-quarters of which are maximally exploited, and many have collapsed.

Global agreements aimed at sustainable fisheries exploitation include the adoption in 1995 of an Agreement on the Conservation and Management of Straddling and Highly Migratory Fish Stocks, and the Code of Conduct for Responsible Fisheries developed by the FAO. Thirty years ago fisheries issues were considered almost entirely in economic and political terms. Today fisheries activities are increasingly recognized as environmental problems in the broader sense.

The global expansion in yields has been delivered by fishing on progressively smaller species at lower levels in the marine food web (the knock-on effects of which are not fully understood) as the top predators have been depleted. The global by-catch of many million tonnes includes not only charismatic animals such as dolphins and turtles but many other species. Effects on marine and coastal ecosystems are poorly known but are probably substantial. Negative ecosystem impacts also result from some types of fishing gear and destructive practices which cause physical damage to the habitat.

While seafood is the primary source of protein for many coastal people, especially the poor, the global demise of fisheries has not been driven only by nutritional needs. Much of the catch is for luxury foods, or is processed into livestock feed. The 'tragedy of the commons' - the absence of a rational reason to restrain harvests that are freely available to all - is one root cause of overfishing while at the other end of the spectrum is so-called 'Malthusian overfishing', when the desperately poor have no choice but to glean the last of the resource.

Many attempts to manage fisheries sustainably have degenerated into a 'division of the spoils'. Political imperatives to maintain employment, international competitiveness or sovereign rights of access have led to fisheries subsidies estimated at up to US$20 billion annually, although these are probably now declining.

The Stockholm Conference and contemporary reports recognized the importance of estuaries and other coastal habitats but the primary concern was the effects of pollution on them. Direct physical alteration and destruction of habitats is now viewed as arguably the most important single threat to the coastal environment. The driving force for physical alteration is ill-planned, and accelerating, social and economic development in coastal areas, which itself results from such increasing pressures as population, urbanization and industrialization, maritime transport and tourism.

Habitat alteration results from activities such as port dredging, landfill, coastal solid waste dumps, coastal construction and road building, the cutting of coastal forests, beach and reef mining, and trampling, anchor and diver damage from tourism and recreation, to name some prominent examples. Failure to consider the economic value of these habitats exacerbates the problem. Mangrove forests, for example, are generally regarded as wastelands ripe for 'reclamation', despite an economic value estimated at around US$10 000/ha/year.

Globally, about one-half of the wetlands and more than one-half of mangrove forests have been lost over the past century, largely because of physical alteration. An estimated 58 per cent of the world's coral reefs are threatened, with direct physical destruction among the most important causes.

Global Warming

The rapid global warming caused by human-induced changes in the atmosphere that is projected by the IPCC would have dramatic effects on the ocean, threatening valuable coastal ecosystems and the economic sectors that depend upon them. Other potential impacts are complex and poorly understood. Polar warming, and melting of the ice caps, could slow down the global atmosphere/ocean 'heat engine', potentially altering the flow of major ocean currents.

The warming of the ocean's surface layers, and an increased input of fresh water, could reduce the upwelling of nutrients that supports much of the ocean's productivity. On the other hand, the highly productive upwelling on the eastern side of some oceans could intensify if, as some projections predict, relatively greater warming occurs there. The IPCC predicts that storms and other extreme weather events will increase in frequency and intensity, increasing natural disturbances to coastal ecosystems and perhaps reducing their ability to recover.

There is particular concern about the possible effects of global warming on coral reefs. During the intense El Niño of 1997-98, extensive coral bleaching occurred on coral reefs worldwide. While some reefs quickly recovered, others, particularly in the Indian Ocean, Southeast Asia and the far western Pacific, suffered significant mortality, in some cases more than 90 per cent. Some models predict a long-term shift to an increased frequency and intensity of El Niño events or similar conditions. If this occurs, bleaching could also become more frequent and intense, with irreversible damage to reefs.

There is evidence that a long-term decline of reefs in the remote Chagos archipelago in the Indian Ocean is related both to El Niño events and to a long-term rise in surface temperature. Mass bleaching of reefs in various parts of the world was also observed in 2000, a possible sign that bleaching is becoming more frequent. Reefs may also be threatened by a higher concentration of CO_2 in seawater which impairs the deposition of their limestone skeletons. Proposed protection measures to address a sealevel rise caused by climate change have shifted away from solid constructions such as seawalls in favour of a mix of soft protection measures (such as beach nourishment and wetland creation), adaptive planning (such as new building codes), and managed retreat, including cessation of new coastal construction.

Some proposals to address global climate change are themselves a cause for concern, particularly those to short-circuit the natural transfer of CO_2 from the atmosphere to the ocean by fertilizing large areas of the ocean surface with nitrogen or iron to enhance phytoplankton growth, or to inject CO_2 directly into deep waters. The effects of these large-scale measures cannot be predicted but are potentially enormous. Small island developing states (SIDS) and lowlying coastal areas are particularly vulnerable to the effects of rising sea levels and more extreme weather.

Furthermore, they are essentially entirely coastal and therefore more dependent upon coastal and marine resources. Recognition of this special vulnerability in Agenda 21 of the UN Conference on Environment and Development (UNCED) led to the adoption in 1994 of the Barbados Programme of Action on the Sustainable Development of Small Island States. Another serious problem is the introduction of marine species to distant habitats where they can multiply uncontrollably, sometimes with devastating effects on the economy and marine biodiversity. Such invasions are occurring around the world with increasing frequency.

The most common medium for species introductions is in ships' ballast water, with about 3,000 species of animals and plants transported every day. Efforts to control species introductions in ships' ballast include the development of new regulations by the International Maritime Organization for ballast water management which is expected to be adopted by 2003.

Africa's 40 000 km of coastline are characterized by a diversity of ecosystems and an abundance of natural resources. The ecosystems include mangrove swamps, estuaries, rocky shores, coastal wetlands and coral reefs; they moderate storm impacts and protect coastal features, recycle nutrients, absorb and break down wastes, provide human and wildlife habitat and maintain biodiversity, and present opportunities for recreation, tourism, transport, trade, and employment. Coastal and marine resources include fish and shellfish, seaweed, wood and fibre, and oil and gas.

Mangrove forests extend from Mauritania to Angola on the west coast and from Somalia to South Africa on the east coast, supporting a diversity of species, many extensively used by local communities. Commercial fisheries contribute significantly to GDP and employment (particularly in small islands). Oil and gas reserves, and other mineral deposits, are also important resources for coastal countries. The growing population and its demands on these resources, however, is causing widespread degradation and pollution of marine and coastal habitats and resources.

Coastal and marine habitats are being physically eroded and biologically degraded through unsustainable rates of resource extraction (including intensive commercial fishing, mining of sand dunes and clearing of mangrove forests). The harvesting methods are also damaging, as in coral extraction and the use of dynamite in fishing. Activities further inland, such as damming of rivers, increased use of fertilizers and clearing of natural vegetation, also affect the coastal zone.

Population growth and migration to the coast, together with rapidly expanding tourism and industrial activities, encourage high rates of infrastructure development, modifying the physical and ecological environment of the coastal zone. Lack of formal protection, sustainable development policies and inadequate resources to implement coastal and marine management have contributed to the pressures, although the situation in many countries is now changing. Coastal drift (erosion and deposition of dunes, beaches and shoreline) is a natural phenomenon but human action can alter natural patterns.

Clearing of forests and natural vegetation inland leads to increased soil erosion and increased sediment load in rivers. Sediment is eventually deposited on the seabed, smothering benthic communities and coral reefs. In contrast, when rivers are dammed upstream, sediment settles before reaching the river mouth, thus depriving coastal zones of sediment. In Western Africa, damming of the Upper Niger, Benue and Volta rivers has altered the flow reaching the Niger Delta, and local subsidence is proceeding at 25 mm per year.

In Ghana, construction of the Akosombo dam in 1965 accelerated coastal erosion west of Accra to 6 metres per year, and in Togo and Benin coastal retreat has exceeded 150 metres over the past 20 years. In Northern Africa, 40-50 per cent of the population in the Mediterranean countries lives in coastal areas, with population densities reaching 500-1 000 inhabitants/km2 along the Nile Delta. In Western Africa, about onethird of the total population is concentrated on a coastal band 60-km wide between Senegal and Cameroon, and large-scale urban growth has occurred from Accra to the Niger Delta, an environmentally sensitive portion of the African coastline.

The coastal zone is also receiving increasing numbers of tourists - in South Africa, for example, the industry grew at 7 per cent a year during the late 1990s. According to FAO (1998), 38 per cent of Africa's coastal ecosystems are under high levels of threat from development-related activities. The exceptional demand for infrastructure development often results in uncoordinated and poorly planned or sited construction which can in turn cause habitat loss, destabilization or mining of dunes for construction materials, and draining of coastal wetlands.

Economic costs are further inflated as governments and investors have to spend large budgets on mitigation and rehabilitation. The demand for fisheries resources is also increasing. The marine fisheries of Africa have developed significantly over the past 30 years, and most demersal stocks are now thought to be fully exploited. The fishery sector contributes more than 5 per cent to GDP in Ghana, Madagascar, Mali, Mauritania, Mozambique, Namibia, Senegal and Seychelles, and the shrimp fishery on the Sofala Bank in Mozambique contributes 40 per cent of the country's foreign exchange.

From 1973 to 1990, fisheries supplied some 20 per cent of the animal protein intake of the population of sub-Saharan Africa. However, per capita fish catch has been fairly static since 1972, except in Southern Africa where it has fallen sharply. The Cape rock lobster and abalone catches have declined steadily since the 1950s, causing concern over the sustainability of these populations and leading to the setting of annual catch limits. In Southern Africa, declining catches, together with a decrease in the mean sizes of fish caught, have led to calls for the protection of line fish stocks.

Today fisheries management measures include minimum size limits, bag limits, use of appropriate fishing gear, closed seasons, control agreements with foreign fleets and establishment of marine reserves. In Western Africa, a Sustainable Fisheries Livelihoods Programme aims to develop social and human capital in fisheriesdependent communities, whilst enhancing natural habitats in those communities.

The waters of the Western Indian Ocean are major sea routes for an estimated 470 million tonnes of oil every year. More than 100 million tonnes of oil are transported annually through the Red Sea alone. This level of shipping incurs a high risk of disastrous oil spills. Furthermore, oil tankers frequently empty ballast and wash engines on the high seas, causing residues of degraded oil to end up on the shore. Port petroleum and oil handling activities also pose threats to the marine and coastal environment.

Accidental leakage from ships, refineries and transport systems are common, especially in Mombasa. Clean-up and disposal of oily wastes is difficult and expensive. Several oil spills off the South African coast have affected African penguins and other marine life. In response, national and regional oil spill contingency plans have been established in several African regions. Effluent from fish processing plants, abattoirs, and chemical and manufacturing industries is frequently discharged into the sea.

In Mozambique, for example, more than 100 factories in and around Maputo do not have waste treatment plants and drain toxic wastes, poisons, non-degradable substances and organic matter into coastal waters. Most of Tanzania's textile mills release dyes, bleaching agents, alkalis and starch directly into Msimbazi Creek in Dar es Salaam. Residues of fertilizers and pesticides washed down in rivers are prevalent in Western Africa, around cities such as Lagos, Abidjan, Conakry and Dakar. Contaminated shellfish can severely reduce economic returns on the catch and may also expose people to gastric and other infections as a result of swimming in contaminated waters or eating the contaminated food.

Domestic solid and liquid waste is also a source of marine and coastal pollution, as municipalities frequently do not have the capacity to deal with the large volumes of waste produced. Solid waste is often dumped on beaches from where it can be blown or washed out to sea. National responses to marine and coastal pollution have included public health legislation and municipal cleaning of coastal areas. International initiatives include the Convention for the Prevention of Pollution from Ships (MARPOL) and the Convention for Cooperation in the Protection and Development of the Marine and Coastal Environment of the West and Central African Region (Abidjan Convention).

However, difficulties have been experienced in monitoring and enforcement, mainly because of the size of the territories requiring policing and a lack of efficient surveillance systems. Other responses have had more success. In Northern Africa, regional emergency plans for containment and clean-up of oil spills have been put in place for the Mediterranean region and the Red Sea. The GEF's US$6 million Industrial Water Pollution Control in the Gulf of Guinea project, which aims to improve the health of the coastal waters between Guinea-Bissau and Gabon, has been instrumental in the adoption of the Accra Declaration, a regional policy for long-term sustainable development in the region.

In the past 30 years, depletion of coastal resources such as fisheries, mangroves and coral reefs has emerged as a critical issue in Asia and the Pacific. Increasing urbanization, industrialization and tourism, coupled with a growing coastal population, have degraded coastal areas, reduced water quality and increased pressures on marine resources. These pressures are exacerbated by poverty. For example, in Viet Nam poor people have become increasingly dependent on marine resources for their livelihoods, and significant beach pollution occurs in the vicinity of Sihanoukville and Kep, both important tourist destinations.

Fish production and aquaculture are practised extensively in the region. Overexploitation of fish stocks and poor aquaculture practices are of concern in Bangladesh, India, Pakistan, Sri Lanka, many Pacific Island countries (PICs) and some other countries. Overexploitation of shrimp resources in coastal waters has reduced exports from capture fisheries and encouraged the growth of aquaculture in almost all countries of the region.

Mangrove clearance for shrimp culture has emerged as a major issue in recent years. It is estimated that more than 60 per cent of Asia's mangroves have already been converted to aquaculture farms. Besides encroaching on mangroves, aquaculture has led to the release of nutrients, pathogens and potentially hazardous chemicals into marine waters. In India, prawn farms have been constructed in low-lying coastal areas, depriving impoverished farmers of agricultural land, causing salinization of groundwater in coastal villages and polluting waterways with excess nutrients.

A number of countries including Australia, India, Maldives, New Zealand, Philippines and Sri Lanka have developed legislation to address problems associated with pollution and overexploitation of fish stocks. Governments have also initiated steps for fisheries management by reducing fishing subsidies and regulating fishing access rights. The South Pacific tuna fishery offers a model of international cooperation for open sea fishing that may prove to be the first sustainable, multinational ocean fishery in the world.

Coral reefs are under stress in many areas, especially those near shallow shelves and dense populations. More than half of the world's coral reefs are located in the PICs, and large areas are already degraded. The causes range from global, large-scale changes in the ocean environment and global warming to tourism and recreation, high population density and economic development in coastal areas since the late 1980s.

Most coral reefs in South Asia were adversely affected by coral bleaching in mid-1998. Extensive damage to reefs has been reported from the Andaman Islands, the Gulf of Mannar in India, Lakshadweep, Maldives, Sri Lanka and the PICs. Increasing water temperatures and increased levels of dissolved carbon dioxide in seawater have resulted in the widespread death of stony corals throughout the tropics.

An important development in coral reef conservation and management was the establishment of the Global Coral Reef Monitoring Network (GCRMN) for South Asia in July 1997 by the International Coral Reef Initiative (ICRI) to facilitate monitoring, training, networking and management of coral reefs. Pollution has considerably degraded the coastal and marine environment, including estuaries, of the region over the past 30 years. Increased wastes from landbased urban, industrial and agricultural activities as well as from offshore oil and gas exploitation are discharged untreated in the coastal region.

The most significant sources of pollution include oil from ships, sewage and other domestic wastes, and industrial effluents. The main route of marine transport of oil from the Gulf is across the Arabian Sea, and accidental oil spills have been frequently reported along oil transport routes, at points of discharge and loading of oil carriers. The shipping of oil coupled with increasing emphasis on offshore oil exploration makes the northern Indian Ocean extremely vulnerable to oil pollution. Oil spills also cause severe pollution in ports in Bangladesh, Indonesia, Malaysia and Pakistan.

In addition, the cleaning of oil tanks in and around ports has led to the frequent formation of tar-balls on the southwestern beaches of Sri Lanka. In the PICs, marine pollution from ships is a threat that is likely to increase as trade and economies develop further. The enhanced use of agrochemicals on land and discharge of chemicals into seawater is a common problem. An estimated 1 800 tonnes of pesticides enter the Bay of Bengal every year. In spite of international regulations, marine pollution in the Sea of Japan and the Yellow Sea has continued to worsen.

Tourism and other recreational activities also pose a threat to coastal ecosystems in many countries. The construction of tourism infrastructure has both a direct and indirect adverse impact on coastal environments through infilling, dredging and resuspension of contaminated silts, discharge of untreated or partially treated sewage, operational leaks, and discharge of hydrocarbons and waste dumping. Sand dunes, an important component of coastal ecosystems in the region, have also been eroded as a result of tourism activities.

Sediment load in the coastal zones of South Asia is high, mainly as a result of soil erosion caused by poor land-use practices and construction activities. Annually, about 1.6 billion tonnes of sediment reach the Indian Ocean from rivers flowing from the Indian sub-continent. The total sediment load of the river system of Bangladesh alone amounts to about 2.5 billion tonnes, of which the Brahmaputra carries 1.7 billion tonnes and the Ganges 0.8 billion tonnes. Coastal erosion is severe in many areas including the Andaman coast, the Gulf of Thailand, Japan and the PICs. The gradual move towards integrated planning and development of coastal and marine areas, through national, regional and global initiatives, is an encouraging trend.

Many countries have adopted the two major international agreements on marine pollution: the London Convention of 1972 and the International Convention for the Prevention of Marine Pollution from Ships (MARPOL) of 1973, with its 1978 Protocol. ESCAP has instituted studies relating to a Coastal Environmental Management Plan for a number of countries in South Asia, including Bangladesh, Pakistan and Sri Lanka. The Plan requires intensive multidisciplinary studies encompassing socioeconomic dynamics, industry, agriculture, fishery, forestry, water resources, energy, ecology and health, as well as close collaboration between the scientific community and governments, other institutions and experts.

Mechanisms for implementing coastal environmental management continue to be developed, and Sri Lanka appears to have made more progress than other countries. Another major multilateral effort that aims at marine and coastal environmental protection at the regional level is UNEP's Regional Seas Programme initiated in 1974. At the sub-regional level the South Asian Seas Action Plan was adopted in 1995 and includes Bangladesh, India, Maldives, Pakistan and Sri Lanka. In 1995, 108 governments across the world adopted the Global Programme of Action for the Protection of the Marine Environment from Landbased Activities. Many countries have also introduced national legislation and projects to address marine pollution.

Europe is almost surrounded by semi-closed and closed seas, such as the Adriatic, Mediterranean, Black, Azov, Caspian, Baltic and White seas. Coastal landscape features range from dunes, cliffs, lagoons and river deltas to very diverse islands, with numerous important marine and bird areas, including 449 Ramsar sites in Western

Europe. The Danube has the largest delta in Europe, encompassing about 580 000 ha.

The limited water exchange of the semi-closed and closed seas with the open ocean makes these seas very sensitive to pollution, which increased dramatically between the 1970s and 1990s, although this trend has been halted and even reversed in a few places in the past ten years. The open coasts of the Atlantic show impacts from land-based pollution, offshore oil and gas, and shipping operations and accidental oil spills. Some 85 per cent of European coasts are at high or moderate risk from development-related pressures.

The rapid development of tourism, increasing transport, intensive agricultural and industrial activities, and continuing urbanization have all put pressures on coastal areas. As a result of infrastructure development and other construction activities, as well as natural causes, coastal erosion is a major issue in some areas, with 25 per cent of the European coast subject to erosion. The challenges for coastal areas are to cope with further economic development and hence growing environmental pressures. Tourism is important for the coastal areas of Europe, considering that they host two-thirds of the region's tourism.

The Mediterranean is the world's leading destination, accounting for 30 per cent of international tourist arrivals and for one-third of the receipts from international tourism. The number of tourists on the Mediterranean coast is expected to rise from 135 million in 1990 to 235-353 million in 2025. Tourism is growing at a rate of 3.7 per cent a year and its demands consume increasing amounts of land. Similar developments can be observed in other important tourist areas along the Baltic, North

Sea and northeast Atlantic coasts. Tourism accounts for 7 per cent of pollution and makes a huge contribution to water scarcity, water consumption by this sector being three to seven times higher than for local populations.

Although shipping is considered to be an environmentally friendly mode of transport, it can have major negative environmental impacts if standards are not observed or enforced. Maritime transport increased in the EU by 35 per cent between 1975 and 1985 but has since levelled off. This has had an impact on SO_2 emissions: maritime transport now accounts for 10-15 per cent of total SO2 emissions. It is estimated that 30 per cent of all merchant shipping and 20 per cent of global oil shipping crosses the Mediterranean every year.

Pollution from land-based sources is still serious in many areas. Many of the 200 nuclear power plants operating throughout Europe are located in coastal regions or along major rivers, because of the large volume of cooling water needed. Since the 1960s, radioactive discharges from the nuclear fleets of the former Soviet navy have affected remote areas of the Arctic and Pacific Oceans. About 150 decommissioned nuclear submarines are rusting in harbours on the Kola Peninsula, Kamchatka and the Russian Far East, representing a potential environmental threat. Although the Helsinki Commission (HELCOM) reports that there is no environmental threat from chemical munitions or radioactive substances in the Baltic marine environment, citizens groups are still concerned.

Discharges from nuclear reprocessing plants originating from the United Kingdom and France are also a matter of concern in the maritime area of the North Sea and the Atlantic. Pollution by heavy metals and persistent organic pollutants, and contamination by microbes and other substances, occur in all European seas. However, there have been some significant improvements. Inputs of hazardous heavy metals and organic substances into the northeast Atlantic fell significantly between 1990 and 1998 after increasing for several decades.

Atmospheric inputs of heavy metals into the North Sea also fell, showing the effect of air pollution abatement policies in the surrounding countries. Between 1985 and 1998, nitrate concentrations decreased by 25 per cent in the coastal areas covered by the Convention for the Protection of the Marine Environment of the North-East Atlantic and the Baltic Marine Environment Protection Commission. The reduced phosphate content of detergents and other measures such as wastewater treatment in catchment areas have resulted in an average decrease of phosphate concentrations in some regions, including the Skagerrak, Kattegat, the German Bight and the Dutch coastal zone.

Wastewater treatment still needs to be improved, however. High population concentrations also result in high levels of wastewater, which is often not sufficiently treated - for example, in the Mediterranean, Adriatic and Black seas. Until the end of the 1980s, large cities on the shores of the Baltic Sea such as St Petersburg and Riga had no wastewater treatment plants. Solid waste is also a problem in some European seas.

A recent study showed that the main sources of solid waste on the coast, sea surface and sea bed in the Mediterranean region are direct disposal from households, tourist facilities and run-off from coastal landfill sites. Global, regional and national measures are being taken to reduce the input of polluting substances into marine waters. International agreements such as OSPAR, HELCOM and the Mediterranean Action Plan (MAP) provide a binding legal framework.

In the OSPAR and Baltic Sea areas, for example, targets have been set to reduce emissions, losses and discharges of hazardous wastes with the ultimate aim of achieving concentrations near background values for naturally occurring substances and close to zero for synthetic substances by 2020. Some states have difficulties in implementing their obligations under these agreements, and this reduces the effectiveness of regional MEAs such as MAP and the Black Sea Convention. Assistance programmes from wealthier states may play an important role in improving implementation and compliance in relation to regional and sub-regional MEAs.

Enforcement has improved significantly in some CEE countries, and the introduction of economic instruments has had an impact. For example, the European Bank for Reconstruction and Development (EBRD) has made funding available for infrastructure improvements in the transition countries in cooperation with HELCOM. However, the slow transformation of large, polluting, state-owned enterprises continues to present obstacles. The recently adopted European Water Framework Directive provides a strong instrument for the control of pollutants and monitoring in the catchment and coastal areas and improvement of water quality for all EU States and incoming accession States.

A recent example of a non-binding agreement at the global level is the Global Programme of Action for the Protection of the Marine Environment from Landbased Activities (GPA). Its implementation will require new forms of collaboration between governments, organizations and institutions concerned with marine and coastal areas at all levels - national, regional and global. Although still in its early stages, the interest and commitment shown by governments in Europe are encouraging.

The main challenge in coastal areas is the implementation of Integrated Coastal Zone Management which aims at harmonizing the various, sometimes conflicting, uses of the coastal zone. In regions such as the Baltic Sea, bordered by several independent nations, transboundary and international cooperation is a basic requirement.

The key environmental problems facing the coastal and marine areas of the Latin American and Caribbean Region are related to habitat conversion and destruction, pollution produced by human activities and overexploitation of fisheries resources. The underlying causes of these problems are linked to the development of coastal areas for tourism, infrastructure and urbanization, and to the conversion of coastal habitats for uses such as agriculture and aquaculture. In addition to diminished natural productivity of coastal areas, most coastal and offshore fisheries are severely overexploited.

These problems are expected to be significantly exacerbated by climate change and sea-level rise, particularly in the Caribbean. The state of coastal areas throughout the region is illustrated in the table on the right. Coastal areas that are heavily populated and exploited require intensive management and infrastructure to sustain coastal ecological systems. However, complicating coastal zone management are multiple physical and political jurisdictions dividing ecological boundaries and scales. The region's coastal zones are the foundation of its economy and sustainability - 60 of the

77 largest towns are on the coasts, and 60 per cent of the population lives within 100 km of the coast.

Development of residential areas and tourism infrastructure has greatly changed the features of coastal areas in the region. Physical alterations of coastlines due to urban growth and the construction of ports and industrial infrastructure are among the major factors that impact the region's coastal and marine ecosystems. Tourism represents around 12 per cent of the GDP of the region, much of which is concentrated along the coasts. Some 100 million tourists visit the Caribbean each year and contribute 43 per cent of GDP and onethird of export revenue. The direct and indirect effects of tourism on coastal and marine areas can be seen in the increasing conversion of coastal habitats and subsequent impacts.

For example, overextraction of ground water by expanding tourism infrastructure results in the intrusion of brackish or salt water into coastal aquifers, eventually contaminating the groundwater system and coastal soil. Pollution is mainly caused by discharge of municipal and industrial solid waste and wastewater, run-off from agricultural fields, and maritime transport, as well as oil and gas extraction, refining and transport. Regional capacity for wastewater treatment is low; some 98 per cent of domestic wastewater is discharged into the northeast Pacific and 90 per cent into the wider Caribbean without treatment.

The effects of pollutants from land-based activities are exacerbated in large watersheds, and in turn may affect distant states. The transboundary effects of five major watersheds are especially notable: the Mississippi, the Amazon, the Plata, the Orinoco and the Santa Marta. Satellite images have shown large sediment discharges from coastal rivers and some large islands travelling across thousands of kilometres of ocean. During a fish kill episode in the Windward Islands in February 2000, pathological bacteria were detected that previously had been reported only in continental freshwater systems. It was suggested that the pathogens had been transported in sediments originating in floods in the Orinoco basin.

Maritime transport is a significant source of coastal and marine pollution in the region especially the release of oil through dumping of bilge water and tank rinsing. Other threats from maritime transport include discharge of sewage, garbage and hazardous chemicals, and introduction of exotic or invasive species to new areas through loading and off-loading of ballast water. The ports in the region are the second most important destination for containerized goods from the United States, and the Panama Canal is a principal link for global maritime trade.

Between 1980 and 1990, maritime transport in the region increased from 3.2 to 3.9 per cent of global trade, and significant increases are expected to continue as a result of trade liberalization and privatization of regional ports. Without counter measures, environmental problems related to maritime transport are expected to worsen in the future. The marine and coastal areas of Latin America and the Caribbean are among the most productive petroleum-producing areas in the world. The most important pressure on the marine and coastal environment in specific localities is the risk of oil spills from oil and gas exploration, production and distribution systems.

The world's largest recorded oil spill was the Ixtoc submarine oil blow-out in the Bay of Campeche, Mexico, on 3 June 1979 with a total estimated outflow of oil greater than the volume from the Exxon Valdez spill. In 1999 and

2001, significant coastal spills and pipeline ruptures in Brazil and Colombia caused both active public concern and new restrictions to control future spills. All oil and gas exploration operations have the potential to cause severe damage to the coastal and marine environment as a result of large and small spills, and chronic leaks. Overexploitation of fisheries resources and the problems of by-catch and discards have become features of the regional fisheries regime.

The catch from the region's seas has generally increased over the past 30 years. Total fish catch reached a regional peak of more than 23 million tonnes in 1994. From 1985 to 1995, many South American countries doubled or tripled their catch, and Colombia's catch increased five-fold. However, in 1998 the regional catch dropped considerably to 11.3 million tonnes, due to adverse climatic factors caused by the El Niño. A recent study that established geographic priorities for marine conservation in the Central Caribbean ecoregion indicated that excessive exploitation was a threat in 34 of the 51 local production systems.

The region also faces the problem of large quantities of by-catch and discards that include turtles, marine mammals, marine birds and other smaller but ecologically important species. At present, the region has no system to record indicators on the health of resources and ecosystems that would gear actions to the recovery of overexploited species and their environment. Measures to halt overexploitation of fisheries have been implemented in some countries.

International policy responses to the problems described above have been many and varied. Most of them are based on fisheries conventions, international shipping conventions, or the large number of agreements tied to the United Nations Convention on the Law of the Sea. At the same time, institutional and organizational weaknesses in the countries of the region, and the myriad authorities responsible for marine and coastal management, make the implementation of policies a difficult task. The Convention on the Protection and Development of the Marine Environment of the Wider Caribbean and its protocols.

UNEP's Regional Seas Programme, and the international project for the elimination of barriers to implement ballast water controls and management measures for developing countries, proposed for the period 2000-2002 by the International Maritime Organization (IMO). The International Coral Reefs Action Network (ICRAN), an important effort to halt the degradation of coral reefs, which is supported by the United Nations Foundation (UNF). The Caribbean Planning for the Adaptation of Global Climate Change (CPACC) project which assists the 12 Caribbean CARICOM countries to prepare for the negative impacts of possible global climate change, especially with respect to the rise in sea level, by measuring their vulnerability and planning for the adaptation and development of their capacity to deal with the problem.

Few of the conventions mentioned, however, have been in force long enough, and with adequate established infrastructure, to assess their strengths and weaknesses. It is clear, however, that regional environmental monitoring processes need to be geared to assessing environmental conditions as well as monitoring implementation activities designed to restore sustainability of coastal and marine areas and their resources.

Almost 25 per cent of Canada's and about 55 per cent of the United States' populations live in coastal areas. The US coastal population is growing at four times the national average, with some of the highest levels of urban growth taking place in small coastal cities. This is of concern because coastal ecosystems are among the richest storehouses of marine biodiversity and provide important ecosystem goods and services.

Conversion of these fragile systems to urban uses can lead to physical degradation, exploitation of marine resources and pollution. Issues of particular concern for the region are the excessive input of nitrogen from land-based activities and the precipitous decline in fisheries: 21 of the 43 ground-fish stocks in Canada's North Atlantic are in decline and nearly one-third of US federally managed fisheries are overfished.

The Pacific Northwest supports rich fishery resources, of which salmon is of primary importance. Historically abundant in many Pacific coastal and interior waters, salmon runs and species diversity have been shrinking since the late 19th century, due to dam construction, rockslides, poor management and overfishing. By the late 1980s, both countries had imposed severe restrictions on harvests of some salmon species but, despite these and other measures, by the early 1990s salmon catch and value showed significant declines; by 1999, 24 subspecies of west coast salmon had been listed under the US Endangered Species Act and Canada had closed or curtailed salmon harvests for some species in a number of its major rivers.

Complicating the issue have been the two international borders that separate British Columbia's waters from Alaska's and those of the northwest United States. During their life cycle, salmon of US origin travel through Canada's waters and vice versa, resulting in a history of intercepting fishery practices that has encouraged unsustainable harvests. The1985 Pacific Salmon Treaty attempted to resolve this issue but broke down in 1992 because of disagreements. A 1999 amendment to the treaty based on sustaining wild stocks, sharing costs and benefits, and a common basis to assess stocks, monitor fish and evaluate performance is more promising.

The combined effects of fishing, climate change and habitat conditions have prompted a number of status reviews, renewed fishing agreements and new management approaches. For example, in 1998 Canada initiated the Pacific Fisheries Adjustment and Rebuilding Program to conserve and rebuild Pacific salmon stocks and to revitalize Pacific salmon fisheries. It has also implemented a precautionary approach to salmon management, resulting in significant harvest reductions to protect stocks at risk.

In December 2000, the United States released a comprehensive, long-term federal strategy to help restore the 14 salmon subspecies in the Columbia River Basin listed on the Endangered Species Act. As those dependent on salmon for income struggle to survive, both countries are taking additional measures to help restore these and other wild fish stocks to the region's coastal and marine waters and to enhance and maintain global biological diversity. Recent restrictions have indeed improved the ocean survival of some important stocks but it remains to be seen if all Pacific salmon species rebound. Nutrient inputs to marine and coastal ecosystems have increased dramatically over the past three decades due to large increases in population density, fossil fuel use, sewage inputs, livestock production and fertilizer use.

These activities release nitrogen and phosphorus, which can enhance plant growth in aquatic systems and lead to oxygen depletion and multiple effects on the ecosystem including destroyed fish habitat, coastal pollution and harmful algal blooms. In many parts of North America, nutrients from non-point sources come mainly from fertilizer and manure run-off. Over the past 30 years, fertilizer use has increased by almost 30 per cent while a trend towards rearing livestock in intensive feedlots has led to the release of large amounts of manure to surface and coastal waters.

Atmospheric inputs of nitrogen derived from manure, as well as from vehicles and electric utility power plants, are also significant. Since the early 1970s, anti-pollution legislation has greatly reduced point sources of nitrogen and phosphorus, principally from the discharge of municipal sewage and industrial wastes and the control of phosphates in laundry detergents. However, most municipal wastewater discharged into Canada's coastal waters is still either untreated or only partially treated. Canadian estuaries in the North Atlantic are less severely affected by nutrient loading than more southerly ones due in part to a cooler climate and significant flushing of coastal waters.

Along the north Atlantic coast, non-point sources of nitrogen are some ninefold greater than inputs from wastewater treatment plants. In 1998, more than 60 per cent of US coastal rivers and bays were moderately to severely degraded by nutrient contamination, and nitrogen was found to be the single greatest environmental threat in some 'trouble' spots on the Atlantic coast. The US Clean Water Act and the 1972 Coastal Zone Management Act directed states to develop management plans for non-point contamination sources and provided funding and incentives to implement them. The 1987 US National Estuary Program aims to minimize regional nutrient contamination.

Nutrient enrichment is probably a contributing factor in the recent dramatic increase in the intensity, frequency and spatial extent of algal blooms or red tides, causing increased economic losses and health impacts. The number of coastal and estuarine locations in the United States with major recurring incidents of Harmful Algal Blooms (HABs) doubled between 1972 and 1995. The impacts of HABs can include human illness and death from eating contaminated fish or shellfish, mass mortalities of wild and farmed fish, and changes in marine food webs.

In response to incidents of human illness from contaminated shellfish, both Canada and the United States have developed testing and water quality programmes to identify phytoplankton toxins and to provide information about them to the public. Ocean acts in both countries establish frameworks for improving the stewardship of North America's coastal and ocean waters.

Since 1996, the North American Commission for Environmental Cooperation has been facilitating regional implementation of the Global Programme of Action for the Protection of the Marine Environment from Land-based Activities in North America. As yet, there is no regional strategy to address nutrient loading in North America's coastal waters, and coordination among the various agencies responsible for their management is inadequate. Evidence suggests that the situation can be reversed, but the need remains for increased political action and changes in the activities in the watersheds and airsheds that feed coastal streams and rivers.

The coastal zones of West Asia are under various degrees of stress as a result of major demographic shifts from rural to coastal urban areas, intense urbanization of coastal zones and dumping of untreated waste. Furthermore, regional wars and internal conflicts have introduced new dimensions to the region's environmental problems and have stressed both financial and natural resources. Actions at national and regional levels, the latter through the Regional Organization for the Conservation of the Marine Environment of the Red Sea and Gulf of Aden (PERSGA countries) and the countries of the Regional Organization for the Protection of the Marine Environment (ROPME countries, under the UNEP Regional Seas Programme's Kuwait Action Plan) focus on the key environmental issues that have emerged in the region: physical alterations, overexploitation of marine resources and marine pollution.

Rapid urbanization has occurred in most of the countries of the region over the past three decades, particularly in the smaller countries such as Bahrain, Iraq, Jordan and Lebanon. By the early 1990s, some of the GCC countries had developed more than 40 per cent of their coastline, and recent estimates indicate coastal investments in the region to be worth US$20-40 million/km of coastline. In Lebanon, more than 60 per cent of the population of about 3.5 million live and work along a very narrow coastal strip. Some 64 per cent of the population of all the Gulf Cooperation Council (GCC) countries except Saudi Arabia live along the western coasts of the Gulf and the Arabian Sea. More than 90 per cent of the population of Bahrain and 37 per cent of Kuwaitis live along the coast.

Coastal populations are expected to increase - for example, the population of Aqaba is predicted to more than double from 65,000 to 150,000 by 2020. Increasing urbanization, accompanied by ill-planned coastal tourism and/or industrial projects, has resulted in the degradation of coastal and marine environmental quality. The Mashriq sub-region and the smaller states of the region are also unable to deal with the large quantities of domestic litter generated along the coasts, due to space limitations and inadequate waste disposal systems. Dredging and land reclamation are also intensifying in most countries.

Major landfilling has occurred along the western coasts of the Gulf countries such as Bahrain, Saudi Arabia and the United Arab Emirates. These activities have led to destruction of marine habitats and ecologically productive areas, coastal erosion, and loss of coastal stretches in many countries. The need for environmental impact assessments and integrated coastal zone management has been recognized by most countries since the early 1990s and a number of coastal and marine action plans have been developed.

A new methodology for integrated coastal zone management was developed by UNEP's Mediterranean Action Plan (MAP), and a Coastal Area Management Project (CAMP) for the south of Lebanon was launched in 2001 by MAP and the Lebanese Ministry of Environment. Nevertheless, with the exception of a regional programme under MAP aimed at safeguarding 100 historical sites notably in the Mashriq region, no concerted effort has been undertaken to protect other historical sites, including submarine structures, from the ravages of dredging and landfill activities. The fisheries of West Asia are diverse and continue to provide protein and revenue. However, per capita fish catches have been falling, albeit slowly, because of adverse climatic and ecological conditions, and unsustainable fishing practices.

The most visible signs of deterioration are the overexploitation and loss of shrimp nursery grounds. Furthermore, fish kill phenomena have often been observed along the shores of the Gulf Area and the Arabian Sea. Fishery regulations are lacking or not enforced, particularly in the Mashriq, and regional cooperation for improved fisheries management is weak. However, a range of policy measures, including the introduction of fishing licences, gear and area restrictions, closed seasons and the banning of certain fisheries, have recently been implemented in the GCC countries.

Initial steps have been taken in several countries to supplement fish protein by aquaculture and/or importation. As aquaculture is expected to increase in both the sub-regions, measures will need to be taken to prevent the accidental introduction of alien species into the wild, which could have adverse impacts on coastal and marine ecosystems. The Mashriq and GCC countries have different sets of pollution-related pressures to contend with. Whereas in the GCC countries the challenges are from oilrelated industries and desalinization plants, in the Mashriq the challenge is primarily from major rivers that discharge domestic and municipal wastes, agricultural chemicals and hazardous industrial substances into the sea.

Due to heavy oil traffic in the Gulf, and the Gulf's unique geographical location and sensitive biological nature, this sea could become the most polluted in the world unless strict measures are implemented and enforced. The Gulf and the Red Sea are the oil tanker highways of the world: more than 10 000 vessels a year pass through the Straits of Hormuz, about 60 per cent of which are oil tankers, and some 34 offshore oil and gas terminals exist in the region. About 1.2 million barrels of oil are spilled in the region every year from routine discharge of ballast water.

Since 1996, facilities for treating oil-contaminated ballast water have been established in the ROPME area, and a task force involving the GCC secretariat, the International Maritime Organization, UNDP, UNEP and the EU was set up through the Marine Emergency Mutual Aid Centre. A regional steering committee was established and a schedule for implementation of oil reception facilities is now under way. More than 360 million tonnes of oil are transported annually in the Mediterranean Sea which, though constituting only 0.7 per cent of the global sea surface, receives 17 per cent of global marine oil pollution.

Around 2,000 vessels, of which 250-300 are oil tankers, cruise the Mediterranean daily. It is estimated that more than 22,000 tonnes of oil entered the Mediterranean during 1987-96 as a result of shipping incidents. Regional wars have also contributed to the degradation of coastal and marine resources. The Iran/Iraq war contributed 2-4 million barrels of spilled oil and 6-8 million barrels were spilled into the Gulf and the Arabian Sea during the Second Gulf War. The region has made some advances in combating accidental oil spills, particularly in the PERSGA and ROPME countries but in the Mashriq countries and some countries in the PERSGA region, there are no mechanisms to deal with major catastrophes.

For example, there are no emergency contingency plans to deal with accidents to the 30-odd oil pipelines in Lebanon. Most countries in the region have recognized pollution from land-based sources as a major threat to the coastal and marine environment. Sewage disposal is a major issue. Most coastal cities in the Mashriq sub-region have outdated sewerage systems, and the discharge of untreated sewage into coastal zones, mainly close to the major cities, continues to be common practice in most of the Mashriq and some parts of the GCC countries.

Elsewhere, as in Bahrain, Kuwait, United Arab Emirates and western Saudi Arabia, all sewage is treated prior to discharge and some is recycled. The risk of eutrophication in semiclosed and closed areas is constantly present since most of the seas in the region are oligotrophic (poor in nutrients). Discharges from desalinization plants of brine, chlorine and heat continue to pose another serious threat to the environment. Nearly 43 per cent of the world's desalinated water is produced in GCC countries and the trend is increasing.

Soil erosion and sedimentation pose another threat to the coastal zone. With an annual estimated soil loss of about 33 and 60 tonnes/ha in Lebanon and Syria respectively, the amount of eroded soil discharged into the Mediterranean from both countries may reach 60 million tonnes annually. In the absence of adequate river basin management programmes, riverine and estuarine water quality will continue to deteriorate with harmful effects on public health.

It is expected that, following the completion of new dams in eastern Turkey, there will be a change in water quantity and quality of the Euphrates River flowing into Syria and Iraq, which in turn will have a major impact on the agricultural areas and estuaries of the Shatt Al-Arab waterways. Although there is great variation in levels of heavy metals in the region, screening tests show acceptable values in most areas. Some countries have begun setting standards for environmental quality through international and regional agreements.

The Arctic marine environment covers approximately 20 million km2 and includes the Arctic Ocean and several adjacent water bodies. Nearly half of the ocean floor is continental shelf, the highest percentage among all oceans. Movements of Arctic waters play a significant role in the global ocean regime, and in regulating the global climate. The Arctic marine environment is rich in fish biodiversity and abundance. The commercial fisheries of the Barents and the Bering systems are among the most productive in the world, with the Bering Sea accounting for 2-5 per cent of the world's fisheries catches. Resident and migratory marine mammals include whales, seals and sea lions. The polar bear is also often classified as a marine mammal because it frequents sea ice in search of prey.

Many of the indigenous communities of the Arctic have traditionally depended on these marine resources for their livelihood. Other natural resources include vast oil and gas reserves along the continental shelves as well as important mineral deposits. However, there are growing concerns about the negative impacts of development activities on the ecology of the Arctic especially in ice-prone areas and critical habitats. Overexploitation of fisheries is a major concern in the Arctic. Since the 1950s, there have been spectacular crashes of populations of commercially important species such as the cod and Atlantic salmon off the coasts of Canada and Greenland, and herring in the Norwegian and Icelandic waters.

Despite strict conservation measures including no-catch zones, recovery has been slow and uncertain. Other species such as the haddock stocks between northern Norway and Svalbard have declined more steadily. Between the 16th and 20th centuries, massive overexploitation of several whale species occurred. Although some species have recovered to sustainable levels, others have not and are still subject to strict domestic or international regulations (for example, the bowhead whale through International Whaling Commission quotas). Illegal exploitation, including of endangered species, and overly generous quotas are a constant threat.

Contaminants are another source of pressure on the Arctic marine environment. The annual spring influx of meltwater carries contaminants which accumulate in the estuaries and deltas and also enter the mixed layer where they are transported to the North American coast. Airborne contaminants from industrial and agricultural activities at lower latitudes are also deposited in the ocean where they can accumulate in sea ice. These contaminants are bioaccumulated in sea mammals and in turn are taken up by Arctic peoples. Radioactive contamination is a further threat, sources including former nuclear weapons testing, the Chernobyl accident, and ocean dumping of radioactive solid wastes which was common until the London Dumping Convention came into effect.

Most of the major changes observed in the Arctic marine environment are believed to be attributable to global warming. For example, the Arctic pack ice is showing noticeable thinning from an average thickness of 3.12 m in the 1960s to 1.8 m in the 1990s. There has been an observed 2.8 per cent/decade negative trend in the ice seasonal cover over the period November 1978 to December 1996. Changes in seasonal patterns of sea ice will affect ocean currents and weather patterns. It is predicted that globally the largest temperature increase will occur in the Arctic.

Arctic countries are taking steps to protect the marine environment. Since the late 1980s they have increasingly engaged in circumpolar cooperation on the marine environment through fora such as the International Arctic Science Committee and the intergovernmental Arctic Council. Cooperative initiatives have included:

— adoption of a Regional Programme of Action for Protection of the Arctic Marine Environment against Land-based Activities in 1998;

— establishment of a trilateral Russian/United States/ Norway Arctic Offshore Oil and Gas Regime that aims to develop a safety and environmental regime for Russian offshore oil and gas operations;

— production of circumpolar guidelines for regulations of offshore oil and gas activities;

— development of a circumpolar protected areas network to include a marine component; and sponsorship of a circumpolar marine workshop with IUCN which developed a set of recommendations to improve the protection and management of the Arctic marine environment.

Given the current warming trend and interest in resource exploitation in the Arctic, it is expected that there will be further exploitation of the Arctic marine environment and increased competition for strategic advantages. However, if the UNCLOS rules for defining limits to resources on the sea bed are applied to the Arctic sea, the wide continental shelves will transfer almost all the Arctic sea bed to national control under the Arctic States. The Southern Ocean represents approximately 10 per cent of the world's oceans. Vast areas of the Southern Ocean are subject to seasonal sea ice which expands from around 4 million km2 in the austral summer to 19 million km2 in the winter.

The extent of Antarctic sea ice has been estimated using Southern Ocean whaling records dating back to 1931. Research suggests a decline in sea ice cover of almost 25 per cent early in this period. However, satellite observations suggest that there has been little change in Antarctic sea ice distribution during the 1970s and 1980s; on the contrary, it seems that the Antarctic sea ice extent increased slightly during these decades. Antarctic fisheries began in the late 1960s with exploitation of the marbled rock cod, a species decimated in the first two years of the fishery. Hydrocarbon contamination of the Southern Ocean is very low and difficult to resolve against natural background levels.

A few spill incidents have been reported in Antarctica during the past decade, the largest of which occurred when the Bahia Paraiso ran aground in the Antarctic Peninsula in 1989, leaking 600 000 litres of fuel. Small diesel spills have been shown to have minor, localized and short-term impacts on the Antarctic marine and coastal environment. However, a large hydrocarbon spill in the proximity of breeding grounds, rookeries or important species habitats could have significant impacts. This is an issue of growing concern as the level of vessel operations, including tourist vessels, in Antarctic waters is expected to increase.

The Antarctic Treaty Consultative Parties have urged those countries which have not yet become Parties to the Protocol on Environmental Protection to the Antarctic Treaty, particularly those with Antarctic tourist activities organized in their territory, to adhere to the environmental provisions of the Protocol as soon as possible. In 1999, the Antarctic Treaty Parties gave priority to the development of safety and environmental guidelines for Antarctic shipping, pending the finalization of the IMO Code of Practice for ships operating in the Polar Regions.

6

Biodiversity

Biodiversity refers to the variability among living organisms from all sources, including terrestrial, marine and other aquatic ecosystems, and the ecological complexes of which they are part. This includes diversity within species (genetic diversity), between species and of ecosystems. Tropical forest ecosystems are the most speciesrich environments. Although th ey cover less than 10 per cent of the world's surface, they may contain 90 per cent of the world's species.

Coral reefs and Mediterranean heathland are also highly species-rich. Around 1.75 million species have been named by taxonomists to date. The total number of species has recently been estimated as 14 million, although this is highly uncertain, due to a lack of information about the number of insect, nematode, bacteria and fungus species. Living organisms contribute to a wide variety of environmental services, such as regulation of the gaseous composition of the atmosphere, protection of coastal zones, regulation of the hydrological cycle and climate, generation and conservation of fertile soils, dispersal and breakdown of wastes, pollination of many crops, and absorption of pollutants.

Many of these services are neither widely recognized nor properly valued in economic terms; however, the combined economic value of 17 ecosystem services has recently been estimated in the range US$16-54 trillion per year. Human health and well-being are directly dependent on biodiversity. For example, 10 of the world's 25 top-selling drugs in 1997 were derived from natural sources. The global market value of pharmaceuticals derived from genetic resources is estimated at US$75 000-150 000 million annually.

Some 75 per cent of the world's population rely for health care on traditional medicines, which are derived directly from natural sources. Biodiversity also provides genetic resources for food and agriculture, and therefore constitutes the biological basis for world food security and support for human livelihoods. A number of wild crop relatives are of great importance to national and global economies. For example, Ethiopian varieties have provided protection from viral pathogens to California's barley crop, worth US$160 million per year. Genetic resistance to disease obtained from wild wheat varieties in Turkey has been valued at US$50 million per year.

Global biodiversity is changing at an unprecedented rate, the most important drivers of this change being land conversion, climate change, pollution, unsustainable harvesting of natural resources and the introduction of exotic species. The relative importance of these drivers differs between ecosystems. For example, land conversion is most intensive in tropical forests and less intensive in temperate, boreal and Arctic regions; atmospheric nitrogen deposition is largest in northern temperate areas close to cities; introduction of exotic species is related to patterns of human activity -those areas remote from human intervention generally receive fewer introduced species.

The ultimate causes of biodiversity loss are human population growth together with unsustainable patterns of consumption, increasing production of waste and pollutants, urban development, international conflict, and continuing inequities in the distribution of wealth and resources. Over the past three decades, decline and extinction of species have emerged as major environmental issues. The current rate of extinction is many times higher than the 'background' rate - that which has prevailed over long periods of geological time.

Information on the conservation status of species is provided by the World Conservation Union (IUCN) which regularly publishes 'Red Lists' of species considered to be threatened with extinction. The latest IUCN Red List indicates that about 24 per cent of mammals and 12 per cent of bird species are currently regarded as globally threatened. Since the Red List assessment in 1996, the number of critically endangered species has increased from 169 to 180 mammals and from 168 to 182 birds.

Insufficient information is available to determine precisely how many species have become extinct in the past three decades. However, the database maintained by the Committee on Recently Extinct Organisms lists 58 fish species and 1 mammal species recorded as extinct since 1970; assessments by BirdLife International indicate that 9 bird species have become extinct during this period.

Table 1
Globally threatened vertebrate species by region

	Mammals	*Birds*	*Reptiles*	*Amphi-bians*	*Fishes*	*Total*
Africa	294	217	47	17	148	723
Asia and pacific	526	523	106	67	247	1469
Europe	82	54	31	10	83	260
Latin America and Caribbean	275	361	77	28	132	873
North America	51	50	27	24	117	269
West Asia	0	24	30	8	9	71
Polar	0	6	7	0	1	14

Much of the relevant information on the status of species is qualitative or anecdotal, and it is therefore difficult to develop a quantitative overview of global trends. To assess trends in species loss or decline, indicators are required that provide quantitative estimates of change over time, using consistent methodologies for sampling and analysis. Ideally, such indicators should be based on data sampled explicitly for this purpose. Few such monitoring programmes have yet been established.

One approach is the Living Planet Index created by UNEP-WCMC in cooperation with WWF. The index is derived from trends in the size of wild populations of species in three habitats - forest, freshwater and marine ecosystems. The prevailing trend of all three indices is downward. The impact of decline or loss of species on the provision of environmental services is difficult to evaluate because the relationship between species diversity and ecosystem function is still unclear.

Some species are known to play more significant roles than others; these have been termed 'keystone' species -

loss of one of these species has a particularly disruptive effect. Reductions in the number of species affect the provision of all ecosystem services because resource capture (of energy, water and nutrients) is greater in more diverse systems. Some ecosystems, such as arid and arctic areas, appear to be particularly vulnerable to human impacts. In these systems, relatively few organisms share common ecological roles. Species diversity may also play a role in buffering ecosystems against the effects of human activity.

The past three decades have been marked by the emergence of a concerted response to the biodiversity crisis. Civil society, particularly in the form of a hugely diverse and increasingly sophisticated NGO network, has been a major driving force behind this. A trend towards increased stakeholder participation relating to conservation action is evident as illustrated by the emergence of partnerships between NGOs, governments and the private sector.

A number of international conventions have been developed that deal specifically with conservation of threatened species. Among the most notable are the 1973 Convention on International Trade in Endangered Species of Wild Fauna and Flora (CITES), and the 1979 Convention on the Conservation of Migratory Species of Wild Animals (CMS) - or the Bonn Convention - which was developed to conserve terrestrial, marine and migratory bird species throughout their range.

Inter-governmental agreements such as the African-Eurasian Waterbird Agreement - which aims to develop transboundary strategic measures necessary to conserve the network of critical wetland areas on which migratory waterbirds depend - are the primary tools for the implementation of the CMS. The focus of conservation action has recently shifted from protecting individual species to conserving habitats and ecosystems. An illustration of how conservation plans are now conceived at broader scales is provided by WWF International which recently developed priorities for action at the scale of ecoregions (large areas of relatively uniform climate that harbour a characteristic set of species and ecological communities).

Ecoregions of particular conservation importance include Lake Baikal in Russia, the Australian Great Barrier Reef and the Atlantic forests of Argentina, Brazil and Paraguay. Loss and degradation of habitat is the most important factor causing loss of species. For example, conversion of forests or grasslands into croplands results in the local extinction of plant and animal species.

Worldwide about 1.2 million km2 of land have been converted to cropland in the past 30 years. In a recent global survey, habitat loss was found to be the principal factor affecting 83 per cent of threatened mammals and 85 per cent of threatened birds (Hilton-Taylor 2000, BirdLife International 2000). Habitat modification arises from many different types of land use change including agricultural development, logging, dam construction, mining and urban development.

Over the past three decades, major losses of virtually every kind of natural habitat have occurred. For example, FAO assessments show that between 1980 and 1995 forest cover in developing countries declined by an estimated 2 million km2 - an average annual loss of 130 000 km2. The most important causes of forest loss included conversion to agriculture and development schemes involving resettlement. As a result, habitats such as the tropical dry forests of Central America have virtually disappeared.

In terms of loss of species, freshwater habitats are the most degraded, with some 20 per cent of freshwater species having become extinct or threatened with extinction in recent decades . The main causes of extinctions among freshwater fishes are declines in habitat quality. Dryland ecosystems, which cover more than onethird of the world's land area, are particularly vulnerable to degradation. Statistics indicate that more than 250 million people are directly affected by desertification. In 1977, 57 million people failed to produce enough food to sustain themselves as a result of land degradation and by 1984 this number had risen to 135 million.

Impacts of degradation on dryland biodiversity have not been comprehensively documented but substantial changes have resulted from grazing of livestock, deforestation, introduction of non-native species and conversion to croplands. In response, the 1977 United Nations Conference on Desertification adopted a Plan of Action to Combat Desertification. As a result the United Nations Convention to Combat Desertification was developed, entering into force in 1996. This convention aims to promote effective action through local programmes and international partnerships.

Wetlands are areas where the water table is at or near the surface of the land, or where the land is covered by shallow water, and include areas of marsh, fen and peatland. Wetlands play an important role in regulating water flow and are of exceptional importance as habitats for large numbers of species. Wetland habitats are also of high economic importance for provision of water and fisheries. Concern about degradation and loss of wetland habitats led to the development of the Convention on

Wetlands of International Importance Especially as Waterfowl Habitat (Ramsar) in 1971.

The Ramsar Convention provides a framework for national action and international cooperation for the conservation and wise use of wetlands and their resources. The designation of protected areas, such as national parks, is one of the most widely used approaches for conserving habitats. In addition to national parks, a total of 167 sites have now been designated as natural heritage sites under the World Heritage Convention. The total area of protected sites has increased continuously during the past three decades from less than 3 million km2 in 1970 to more than 12 million km2 by the late 1990s, indicating that there are continuing efforts by governments to establish protected areas.

Although the effectiveness of protected areas for conserving biodiversity has been questioned, a recent analysis of 93 protected areas around the world indicated that most parks are successful at stopping land clearing and to a lesser extent at mitigating logging, hunting, fire and grazing. The most significant response to the biodiversity crisis during the past 30 years has been the Convention on Biological Diversity (CBD) which entered into force in December 1993 and had been signed by 182 Parties by December 2001. The convention has three main goals: the conservation of biodiversity; sustainable use of the components of biodiversity; and sharing the benefits arising from the use of genetic resources in a fair and equitable way.

The CBD has resulted in major activity at both national and international levels, and in the increased coordination of cross-sectoral action within and between countries. However, major challenges remain in increasing capacity to assess biodiversity and its value to people, securing adequate financial resources for conservation actions, and building political support for the changes necessary to ensure biodiversity conservation and sustainable use.

Climate Change and Global Warming

During the 1990s climate change emerged as one of the major potential threats to biodiversity. The IPCC concluded that climate change could lead to severe adverse impacts on ecosystems, and on the goods and services they provide. Some ecosystems might disappear, while others could experience dramatic changes in species composition. Desertification may increase in some areas and some species could also become more vulnerable to extinction. The impact of climate change on biodiversity to date is unclear. The increasing incidence of coral reef bleaching may be a consequence of recent rises in global ocean temperature.

The most significant mass bleaching was associated with the 1997-98 ENSO event, when all ten reef provinces of the world were affected. In some areas, most notably the Indian Ocean, this event was followed by mass mortality, where up to 90 per cent of all the corals died over thousands of square kilometres. Climate change has also been implicated in the decline of amphibians in tropical montane forests.

Nitrogen deposition

Nitrogen deposition has become a major cause of biodiversity loss. It has increased substantially in recent decades, primarily as a result of an increase in the use of fertilizer and the burning of fossil fuels. Increased nitrogen in soil and water can lead to loss of species and shifts in the species composition of plant communities; for example, the conversion of heathlands to species-poor grasslands in the Netherlands. Aquatic ecosystems are the most vulnerable; nitrogen deposition can lead to eutrophication, currently one of the most serious threats to aquatic environments, particularly in inshore waters where many commercial fish and shellfish species breed. Nitrogen deposition has also been associated with the recent increase in toxic algal blooms.

Oil spills

Oil spills have also had a major impact on biodiversity in recent decades. In 1998 alone, a total of 108 000 tonnes of oil were spilled worldwide into marine and inland environments as a result of 215 incidents.

Consumption and international trade

During the past 30 years, consumption of natural resources has increased substantially - for example, the global consumption of forest products such as paper increased threefold. For many biological resources, such patterns of increasing consumption are unsustainable. The most striking example is marine fisheries. Consumption of fish has increased 240 per cent since 1960. However, the marine catch has now levelled off and shows signs of declining as a result of overexploitation.

More than 70 per cent of the world's commercially important fish stocks are described by FAO as either fully fished, overexploited, depleted or slowly recovering. Numerous fisheries collapsed during the latter part of the 20th century, including Canada's Grand Banks cod fishery which closed in 1992 with the loss of 40 000 jobs. Products derived from wildlife form the basis of an international trade valued at approximately US$10 000 million annually. Additionally, there is an extensive illegal trade in such products.

In addition to the CITES secretariat, the establishment in 1976 of the Trade Records Analysis for Flora and Fauna in International Commerce by IUCN and WWF has strengthened the international community's efforts to monitor illegal wildlife trade and implement CITES' provisions and decisions. The CITES secretariat, Interpol and the World Customs Organization, as well as a number of NGOs, have been establishing networks and organizing training of customs, borders, police, wildlife and other enforcement authorities.

The impact of CITES on biodiversity is difficult to assess, as it is often not possible to ascribe unequivocally any changes in the conservation status of species to the impacts of actions taken under the convention. The continual upgrading of many species to increasingly higher levels of protection suggests ineffectiveness, although some species (such as vicuña) have been downgraded because of successful sustainable use schemes.

Invasive species

Invasive species are organisms (usually transported by humans) that successfully colonize native ecosystems. Such species have been a major threat to native species through the effects of predation, alteration of habitat or disruption of ecosystem processes. Notable terrestrial examples include the loss of many endemic land snail species of French Polynesia following the introduction of the predatory snail Euglandina rosea, and the decline in New Zealand's native birds due to the introduction of Australian brushtail possum. Aquatic examples include the introduction of the predatory Nile perch Lates niloticus to Lake Victoria around 30 years ago, which contributed to the apparent extinction of 250 endemic species of cichlid fishes.

The number of aquatic introductions rose rapidly during the second half of the 20th century. The CBD recognizes the importance of invasive species as a global problem and calls upon contracting parties to prevent the introduction of, control or eradicate those alien species that threaten ecosystems, habitats and species. In response to a recommendation from the CBD in 1996, the Global Invasive Species Programme (GISP) was developed, which is coordinated by the Scientific Committee on Problems of the Environment (SCOPE), in collaboration with IUCN, Centre for Agriculture and Biosciences International and UNEP.

Biotechnology

Biotechnology is increasingly being used for the genetic improvement of crops but concerns have been raised about potential risks to biodiversity. The organisms produced are referred to as genetically modified organisms (GMOs) or living modified organisms (LMOs) and efforts have focused on such crops as tomatoes, grains, cassava, maize and soybeans. In response to this concern, a subsidiary agreement to the CBD was negotiated to address the potential risks posed by cross-border trade and accidental releases of GMOs.

Five internationally recognized 'biodiversity hot spots' (areas of particularly high species richness and endemism, and under particular threat) are found in the African region. These are the Western Indian Ocean islands, the Cape floristic region, the Succulent Karoo (the most species-rich desert in the world), the Upper Guinea forest and the Eastern Arc mountain forests of Eastern Africa. Part of the Mediterranean Basin hot spot, home to 25 000 plant species and 14 endemic genera, is also found in Africa.

The continent possesses several other areas of great importance for biodiversity. These include the highlands of Ethiopia; the forests of the Albertine Rift in Burundi, eastern Congo, Rwanda, and adjacent parts of Kenya and Uganda; the western escarpment of Angola; and the miombo woodlands of interior Southern Africa. In the past three decades, habitat loss and degradation has been a major issue throughout Africa, particularly in dryland areas.

In humid areas, the bushmeat trade has also had a significant impact on biodiversity. Biodiversity resources are extensively used for subsistence and commercial purposes. For example, approximately 70 per cent of the wild plant species in Northern Africa are used as sources of traditional food, forage, medicine and agroforestry, and half have more than one use. The richness and diversity of ecosystems in Africa underpins a flourishing tourism industry; which is an important source of foreign exchange for many countries. For instance, Southern Africa's wildlife attracted more than 9 million visitors in 1997 bringing in a total of US$4.1 billion.

FAO's Global Forest Resources Assessment estimated the rate of deforestation in Africa during the period 1990-2000 as 0.78 per cent of total forest area a year, representing a yearly loss of some 5.2 million ha. The cause is primarily clearance for agriculture but extraction of timber and fuelwood, fire and overgrazing have also been important factors. Deliberate burning of grasslands is widely practised in many African countries, with 25-50 per cent of land cover in the arid Sudan zone and 60-80 per cent in the humid Guinea zone burned annually.

Impacts of habitat loss and degradation on biodiversity are difficult to evaluate. However, dramatic contractions in the range of many species have been recorded. For example, in Africa as a whole, elephants declined from about 1 300 000 to 500 000 during the 1980s. Declines were most pronounced in areas characterized by poaching, civil war, high rates of land use change and increases in human population densities.

Central Africa had lost about half of its wildlife habitats by 1986. Draining of wetlands for agricultural and urban development, degradation through overgrazing and collection of fuelwood, and pollution through effluent discharge have caused the loss of up to 50 per cent of wetlands in Southern Africa and Western Africa, while some 80 per cent of the Upper Guinea forest has now been cleared. During 1980-95, the number of recorded extinct plants in Southern Africa increased from 39 to 58, and the number of threatened plants more than doubled.

The main response to loss of natural habitat has been the establishment and extension of protected areas. Overall, approximately 7 per cent of the land area of Africa has been designated as protected. In total, Africa contains 1 254 protected areas, including 198 marine protected areas, 50 biosphere reserves, 80 Wetlands of International Importance and 34 World Heritage sites. Protected area coverage differs markedly within Africa; for example, a substantially higher proportion of the land area is designated as protected in Southern Africa than in other sub-regions.

Lack of financial support and weak law enforcement are common problems in African protected areas, resulting in encroachment by human activities and settlements. However, sub-Saharan Africa accounts for 18 per cent of the global mean investment in protected areas. Protected areas are being increasingly managed for multiple uses, including tourism and sport hunting. Some 52 African countries are party to the Convention on Biological Diversity, 48 countries are party to CITES, and 22 are party to CMS. This is reflected at the national level in the development of national action plans and strategies for the environment, biodiversity and conservation.

Financial assistance from a range of bilateral and multilateral donors offers opportunities to address the key issues relating to biodiversity and to promote sub-regional cooperation in conservation. Several transfrontier reserves are being established in Southern and Eastern Africa. During the colonial era, conservation policies were often based on protectionism that ignored the needs of African people, by imposing hunting restrictions and excluding people from reserves. Protected areas fell under this category and have been described as 'fortress conservation'.

Policies on wildlife conservation have since changed with communities living adjacent to national parks being considered as partners; a key trend during the past three decades has been the increasing involvement of local people in conservation initiatives. Community-based conservation (CBC) programmes seek to achieve this by allowing people living near protected areas to participate in land management decisions, giving people rights to wildlife resources and ensuring that local people derive economic benefit from wildlife conservation.

Impacts of wild harvest

In much of sub-Saharan Africa, the harvest of wildlife for food has a major impact on the populations of many species. Wild food may play an important role in food security for rural people and is also, increasingly, a commercial commodity that is traded nationally and regionally. In many urban areas, meat from wild animals commands a significantly higher price than that from domestic animals, helping to stimulate largescale harvest. Large quantities of meat are involved: in the Central Africa moist forests alone as much as 1 million tonnes of wildlife are killed for food each year. A great deal of the wildlife harvest in Africa is believed to be currently unsustainable and has been implicated in the declines and local extinctions of a range of animal species.

A number of wild plant species are affected by harvest for medicinal purposes. Rural and urban populations all over Africa depend largely on medicinal plants, often collected in the wild, for their health needs. Some species, such as the montane tree Prunus africana and the southern African devil's claw Harpagophytum species, are also exported in significant quantities.

Overharvesting, together with agricultural encroachment and unregulated burning, are believed to be contributing to the decline of many species in the wild. In a survey of medicinal plant use in 17 countries in East and Southern Africa, more than 100 indigenous plant species were identified as conservation or management priorities on a national basis. Over the past 30 years, trade regulation and prohibitions or suspension, mainly through CITES, have been used to control international trade in threatened species with varying degrees of success.

Indonesia is thought to support more species, with more endemic species, than any other country in the world, closely followed by several others, including Australia and China. The tropical waters around the Indo-Australasian archipelago are the world's centre of diversity for a wide range of marine groups, including corals, coral reef fishes and mangroves. Rangelands in western parts of the region, the Tibet plateau and Australia are particularly rich in lizards and snakes adapted to arid conditions.

Many of the rivers and freshwater lakes hold endemic species of fish and aquatic invertebrates. The larger islands are home to a wide range of endemic species while the continental areas often have high species richness together with high rates of endemism. Such 'hot spots' can be identified at a range of scales, from individual mountains to extensive hill ranges. The entire Hindu Kush-Himalayan belt has as many as 25,000 plant species, comprising 10 per cent of the world's flora. A few such areas remain relatively unknown: remarkably, even new large mammal species have recently been described in Viet Nam and Laos.

Biological resources have long been of subsistence importance, and have been increasingly exploited for trade. At the global level, around three-quarters of known or suspected species extinctions have occurred on isolated islands, many of which were molluscs and birds from the

Asia-Pacific region. Some 1,469 vertebrate species in the region are currently considered to be threatened with extinction. Habitat loss is the principal factor that fragments natural populations and increases their risk of extinction but this often acts in synergy with other pressures such as alien species and unsustainable harvesting.

Alien species

Introduced species have long been recognized as a threat to indigenous species, particularly species endemic to single countries or small islands. For example, native plants on the main islands of New Zealand compete with a range of introduced plants and are heavily affected by introduced terrestrial mammals, among which brush-tailed possums (from Australia) are a particular threat. Tens of millions of New Zealand dollars were spent annually on possum control in the 1990s to reduce habitat loss and control bovine tuberculosis which can be passed from possums to domestic cattle. New Zealand birds, reptiles and amphibians are also under pressure from introduced predators such as stoats, rats and cats but much emphasis is now given to invasive species control programmes on small islands, where long-term control may be feasible. The robin Petroica traversi was formerly widespread in the Chatham Islands but had been much reduced by the late 19th century.

By the 1970s, the species was restricted to Little Mangere Island, where the remaining patch of forest was being destroyed by invasive plants. A conservation programme has now resulted in a population of some 200 birds, all descended from a single pair. The brown tree snake Boiga irregularis spread widely through Guam, from the 1950s onward, after being accidentally introduced in military aircraft. It has had a severe impact on the native bird fauna, one species of which is believed to be extinct, one is extinct in the wild and one has been assessed as critically endangered. Molluscs in Moorea provide a striking example of the potential impacts of introduced species.

A carnivorous snail from Florida, Euglandina rosea, was introduced to control numbers of the giant African land snail Achatina fulica that had become an agricultural pest after itself being introduced to the island. The introduced carnivore subsequently preyed heavily on the endemic native snails in the genus Partula, all seven species of which are now extinct in the wild -although they survive in captivity.

The natural vegetation over much of the region is forest, with grassland, scrub and semi-desert occurring in the more arid regions. Much of the forest cover has been recently cleared by humans so that temperate forests are reduced in China, Japan and New Zealand, while tropical forests are similarly reduced in South Asia and Southeast Asia. More extensive forest cover persists in Borneo, Myanmar and New Guinea. Timber and non-timber forest products (rattan, bamboo, resins, waxes, nuts, honey, spices and medicinal plants) provide a livelihood for indigenous people. Timber extraction and clearance for commercial plantation have caused much of the forest loss, reinforced by increasing human numbers and development pressures.

National policies and plans for management of forest resources exist and there is an increasing amount of plantation forest but serious decline in natural forest cover has occurred in many countries. Indonesia recorded an average annual decrease between 1990 and 2000 of 1.3 million ha, one of the highest rates of deforestation recorded globally. Malaysia, Myanmar and Thailand all also show major decreases of 237 000, 517 000 and 112 000 ha respectively, with equivalent percentage losses of 1.2, 1.4 and 0.7. This trend is of major concern. If current trends continue, Indonesia's lowland forests will be destroyed by 2005 on Sumatra and by 2010 on Kalimantan.

Although the benefits of dams can be substantial, their negative impacts, including significant biodiversity loss, are common. Analysis by the World Commission on Dams has shown that the environmental, hydrological and economic arguments used to support dam construction are often flawed. Impacts on biodiversity are not well documented but major river systems are drying, groundwater is increasingly overexploited and pollution is a major concern; as a result biodiversity impacts are likely to be substantial.

The fish yield from the reservoir was far below the level predicted in the 1981 impact assessment and yields from the original free-flowing river were underestimated. Some 50 fish species dependent on rapids have disappeared and migratory fishes have declined; these biodiversity losses have had serious impacts on households that depend on fishing. The lack of evaluation of the likely impacts on fishes and fisheries is identified as a critical omission from the original impact study. More comprehensive impact assessment, better evaluation of biodiversity impacts and greater emphasis on mitigation and remedial measures will be needed for future dams.

In response to deteriorating biodiversity, many countries are party to international agreements. All except

Afghanistan, Brunei Darussalam and Thailand are party to the Convention on Biological Diversity (CBD). The CBD specifies a framework of measures at the national level for the conservation of biodiversity and many countries have prepared National Biodiversity Strategy and Action Plans and submitted National Reports. Most countries are party to CITES and the Ramsar Convention on Wetlands.

National responses aimed at conserving biodiversity have been variable in effectiveness, with many initiatives suffering from a lack of data and common understanding of ecological systems. Protected areas have been set up in different countries but they tend to be geographically limited and disconnected. The proportion of protected area to total area in most countries is lower than the 10 per cent norm recommended by IUCN.

Europe is home to a wide variety of ecosystems, ranging from the Atlantic coast to the Russian steppes, and from the boreal forest and tundra of Scandinavia to Mediterranean forests and shrubland. Europe is also an important crossroads for large populations of migratory species shared with Africa, West Asia and North America.

Agricultural land covers some 45 per cent of Europe and most natural habitats are therefore restricted in extent. The impact of agriculture on biodiversity is thus a key issue. The genetic modification of organisms for agriculture has also emerged as an important issue relating to biodiversity. The landscape has been significantly modified by human activities, including deforestation, agriculture, drainage of wetlands, modifications to coastlines and river courses, mining, road construction and urban development. As a result, natural habitats have been reduced in size and fragmented, and are therefore less able to support wildlife. Habitats such as lowland forests and wetlands have undergone particularly large declines. Relatively pristine areas remain in some Nordic and Eastern European countries.

The direct impacts of agriculture include effects on water quality, land drainage, soil erosion, toxic effects of fertilizers and biocides, and the destruction, degradation and fragmentation of habitat. This has had a substantial negative impact on biodiversity, and population declines and range contractions have been found to be significantly greater in countries with more intensive agriculture. In the United Kingdom, 26 species of farmland birds declined significantly during 1968-95, primarily as a result of agricultural intensification.

Intensive agriculture also often leads to the eutrophication of freshwater habitats, resulting in deoxygenation of water, production of toxins and a general decline in wildlife conservation value. Some 46 per cent of Ramsar lake sites in Europe have suffered a decline in water quality, largely as a result of eutrophication. Wetland habitats have also been affected by land reclamation for agriculture. In Spain alone, more than 60 per cent of all inland freshwater wetlands disappeared during a 25-year period.

Nutrient enrichment also has a significant impact on marine ecosystems, particularly in areas such as the Caspian Sea. Such pollution has increased the frequency of algal blooms in the Adriatic, where they foul fishing gear and beaches, and in the North Sea where in 1988 they caused massive mortalities of farmed salmonid fish. Key trends over the past three decades include the agricultural 'improvement' of low-intensity farmland, which has led to substantial and mostly irreversible loss of habitats through drainage, fertilization and increased stocking densities.

In addition, hedgerows, field verges and grass tracks have been lost as a result of increased field sizes and mechanization. During the 1970s and 1980s, some 27 200 km of hedgerows were lost annually in England and Wales. As a result of agricultural mechanization, much of the remaining area of scrub and grassland habitat is now restricted to ground of low agricultural value, such as steep slopes and poor soils. Agricultural intensification has also resulted in the loss of fallow land and stubble, important habitats for wildlife, particularly birds.

The importance of agriculture to biodiversity has been recognized in several policy responses. A European Commission Biodiversity Action Plan for Agriculture has been developed as part of EC commitments to the Convention for Biological Diversity (CBD) through the 1998 Biodiversity Strategy. The plan is designed to help integrate biodiversity targets into relevant policy sectors. Strategic questions relating to agricultural policy are mainly addressed within the Common Agricultural Policy (CAP) and a key element of the Biodiversity Strategy is therefore to incorporate biodiversity objectives into CAP.

In Western Europe, more than 22 million ha of agricultural land are covered by some form of agreement to maintain biodiversity and landscapes. This exceeds the target set in the EU's fifth environmental action programme. However, the extent varies - from more than 60 per cent of farms in Austria, Finland and Sweden, to 7 per cent or less in Belgium, Greece, Italy and Spain. The environmental performance of these schemes is open to doubt since many lack precise objectives and have no monitoring provisions.

Since the early 1980s, grassland and heathland habitats have benefited from a shift in agricultural policies.

For example, in Germany and Italy the reform of EU policy has released more than 300 000 ha of arable land for conversion to grassland in lowland areas, as part of the 'set-aside' programme. While initially welcomed as an opportunity to increase the ecological values of such areas, 'set-aside' measures can also have negative results - causing people to abandon traditional farming systems and adopt inappropriate forms of forestry or afforestation.

Protection Measures

Only 5 per cent of the land area of Europe is currently designated as a protected area. The major policy instruments relating to habitat protection are Agenda 2000, Natura 2000, the Emerald Network and the Pan-European Ecological Network. With these it is planned to create a coherent European ecological network of natural and semi-natural habitats and provide or restore corridors between existing protected areas throughout the region. Agenda 2000 is an action programme designed to strengthen EU policies. The programme will promote new interrelationships between rural areas and biodiversity, involving agri-environmental measures, structural funds, Less Favoured Area measures, afforestation measures, and so on.

For non-EU countries a less binding programme (the Emerald Network) was set up recently under the Bern Convention. Some eastern European countries have already established Natura 2000 networks. These developments are key elements in Europe's contribution to the CBD. EU strategy aims to complement biodiversity initiatives at the national level through a series of action plans to integrate biodiversity into other sectoral policies and programmes. Similarly, national biodiversity action plans are being developed throughout much of Europe.

Countries in Central and Eastern Europe still possess a wealth of well-preserved landscapes, ecosystems and species that are rare or already extinct in Western Europe. Most protected areas in these areas had been designated by the end of the 1970s, often surrounded by large buffer zones and connected by habitat corridors linking sites. However, with economic transition, the system of nature protection came under intense pressure as state financing declined and it is now in jeopardy.

Genetically modified organism (GMO)

Genetically modified organism (GMO) technology could play an important role in increasing agricultural production in Europe. However, the release of GMOs into the environment remains a subject of controversy. Experimental releases of GM crops have been conducted in both Western and Eastern Europe but experience with actual cultivation is still limited. In Western Europe, the public is generally sceptical about GM foods and organisms. There is strong support for labelling, public consultation and more comprehensive regulation and monitoring. Apart from food safety, concerns have also been expressed about adverse effects on the environment and biodiversity; for example, genetic transfers with native species.

While developers of GMOs see a huge market opportunity, food producers are under pressure from consumers who wish to avoid GMOs. Efforts in Western Europe are currently aiming to inform, involve and consult the public about GMOs, in order to reach a consensus on regulation. In Eastern Europe some NGOs are trying to bring the issue into the open and a regional biosafety process, started in 1995 in Hungary, is also furthering discussion. The European Commission has proposed new legislation to harmonize action and to facilitate agreement on marketing authorization.

Current European legislation is consistent with the Biosafety Protocol that was adopted in January 2000 as part of the CBD. The potential risks of GMOs to biodiversity are the subject of ongoing research. Public awareness also needs to be raised to ensure a well-informed and balanced multi-stakeholder dialogue and decision process.

The region contains a wide variety of ecosystem types. Tropical moist and dry broadleaf forests cover 43 per cent of the territory; grasslands and savannas 40.5 per cent; deserts and scrub 11 per cent; temperate forests and tropical and sub-tropical coniferous forests 5 per cent; and mangroves the remaining 0.5 per cent. The region's rivers and lake ecosystems and the marine ecosystems of the Pacific and Atlantic coasts are also productive habitats with high diversity of species. The Caribbean contains 7 per cent of the world's coral reefs with a great array of marine biodiversity.

Cloud forests and other humid montane forests have been identified as one of the most threatened habitat types in the region. They are found where persistent cloud cover is in contact with the mountainside, at altitudes of 1 000-3 000 metres and play a critical role in the provision of clean water supplies to human populations in the lowlands. Humid montane forests also harbour the wild relatives and gene pools of many New World crops, including potatoes, maize and beans.

The principal pressures on cloud forests are clearance for subsistence and commercial agriculture by rural communities, and in some regions for plantations of narcotics. Human population growth and poverty drive these processes but the construction of roads and increased links to commercial markets has also stimulated the production of cash crops. Other major pressures include deforestation for cattle ranching, which in the past has often been supported by government policies.

Lowland tropical rainforests have been the focus of particular conservation concern, being the habitat with the highest species richness and with continuing conversion of major areas to other land uses. The Brazilian Amazon is the largest tropical rainforest in the world, which once had a forested area of 4 million km2. By 1998, 86.3 per cent of this area remained, with 377 200 km2 cleared during the preceding 20 years. The mean rate of forest clearing accelerated in the 1990s, and the total area affected by fragmentation, clearing and edge effects is now estimated to comprise one-third of the Brazilian Amazon.

Deforestation in the Brazilian Amazon is driven by several processes. A major pressure is the ten-fold population growth in the region since 1960. In addition, industrial logging and mining and their associated networks of roads, which make accessible new areas of forest to colonizers and ranchers, have been major factors in deforestation. Some 6 per cent of the region falls within the strictly protected category. Major efforts have been made in the field of timber certification and forest preservation to reverse the loss of biodiversity. Human-ignited fires have also become widespread, especially in logged and fragmented areas.

The forests of eastern coastal Brazil are considered among the most endangered habitats on earth and have been given highest priority for biodiversity conservation. They contain 7 000 endemic plants and 779 endemic vertebrates - 2.7 and 2.1 per cent of the global total, respectively. In the Bahia region, only 0.4 per cent of continuous forest cover remains of the original forest area of 215 436 km2. Threats arise from coastal development and uncontrolled logging, agriculture and charcoal production.

Overall, more than 10 per cent of the region is currently protected. In addition, the attractiveness of the cloud forests and recognition of their values by concerned individuals has led to the creation of many private forest reserves in the region, often linked to scientific research programmes and ecotourism ventures. A new approach to promoting montane forest conservation in the region is to compensate forest owners for the environmental services their forests provide to society, often financed by the collection of a small surcharge on the users of water originating in the forests. Such schemes are being considered in several Latin American countries and have been tested in Costa Rica.

Many forest conservation initiatives have been developed in the Amazon, including land-use planning, the establishment of protected areas linked by corridors and extractive and Amerindian reserves. The largest of these is the Pilot Programme to Conserve the Brazilian Rainforest, with support from the G-7 nations. However, further major infrastructure, industrial agriculture, mining and logging projects are currently planned for the Amazon. The CBD has played an important role in terms of the response to biodiversity loss. While some countries have incorporated the CBD's objectives in general legislation, others have done so by means of sectoral laws.

The former group includes Brazil, Colombia, Costa Rica, Peru and Venezuela. For example, Brazil established a National Programme on Biological Diversity in 1994 together with an accompanying project for the conservation and sustainable use of Brazilian biological diversity (PROBIO), identifying priority conservation areas and actions through a series of assessments. In Peru, the Law for the Conservation and Sustainable Use of Biological Diversity, which covers most CBD commitments, entered into force in 1997.

It is expected that the nine Caribbean countries now preparing national strategies on biodiversity will implement the CBD by means of legislation, development of institutional mechanisms and by providing adequate resources. Countries that are modifying their sectoral laws include Cuba, Honduras, Mexico, Nicaragua and Panama. However, legislation for the implementation of the CBD has often been developed without reference to other biodiversityrelated conventions such as CITES, the Convention on Migratory Species and Ramsar.

National funding programmes such as the Mexican Fund for Nature Conservation have been established as part of national efforts to implement the CBD. Additional sources of funding include organizations such as the World Bank and the Inter-American Development Bank, together with other international agencies, NGOs and bilateral cooperation agencies. Between 1988 and 1999, the World Bank Group approved 74 biodiversity projects in the region which were declared consistent with the goals and objectives of the CBD. A substantial amount has been distributed among regional efforts to conserve biodiversity, especially since 1995. As expected, most of the resources went to the largest countries. Brazil alone received 56 per cent of the total but this benefit has not been equally distributed among ecosystems, the majority going to the Amazon and Atlantic rainforests.

Unsustainable harvesting and illegal trade

Illegal trade in plants and animals is one of the greatest threats to biodiversity in many countries including Brazil, Colombia, Mexico and Peru. It is difficult to measure the extent of this illegal trade and its impact on lesser-known species. Estimates suggest that Brazil accounts for 10 per cent of the global wildlife trade, which is valued at approximately US$10,000 million per year. Despite ongoing efforts, including development and implementation of national strategies to control illegal trafficking in countries such as Colombia, police records on seizures confirm that illegal trade of flora and fauna remains a widespread problem.

National governments are responding to this issue in a number of ways. For example in Colombia, the sale of some wild animals is permitted for domestic and international markets. There are 50 private establishments with legal authorization to capture caiman, iguana, boa, black tegu and capybara for processing and marketing. As a result, in the year 2000, 739,000 caimans, 232,000 iguanas, 3,530 boas, 2,700 black tegu and 10,000 capybaras were captured for market in accordance with national regulations and the recommendations of CITES.

Habitat destruction and degradation is the most pervasive threat to biodiversity in the region. North American wetlands have high biological productivity, providing critical habitats for many species and essential ecological services such as taking up floodwaters and protecting water quality by filtering pollutants. Wetland protection is therefore a priority issue for biodiversity conservation in North America. Another key issue is the threat that non-native species pose to native species through predation, competition, parasitism and hybridization.

North America contains many different ecosystems, with biodiversity increasing along a northsouth gradient and the Hawaiian Islands containing the highest diversity of species. North America contains a large percentage of the world's wetlands with Canada holding about 24 per cent, accounting for about 16 per cent of its landscape. Wetlands cover about 264 million ha of North America. According to Canada's endangered species list, as of May 2001 a total of 352 species were at risk of imminent or eventual extinction while in the United States 1,231 species were listed as endangered or threatened.

Some 309 vertebrate species are threatened with extinction in the region. To safeguard biological diversity, North America has set aside protected areas. More than 14 per cent of the region's land area is now protected, with 4,521 protected sites covering an area of some 264 million ha. Canada has signed and ratified the Convention on Biological Diversity (CBD) and continues to work towards introducing a federal Species At Risk Act. The United States is not yet party to the CBD but has a strong Endangered Species Act. The latter has been used effectively by NGOs to protect substantial areas of habitat for threatened species.

Wetlands

Wetlands provide food and habitat for about one-third of bird species in the United States and more than 200 species in Canada. They are also home to some 5 000 plant species and 190 kinds of amphibians in the United States and 50 species of mammals and 45 species of waterfowl in Canada. About one-third of North America's threatened and endangered species live in wetlands. Prior to the 1970s, government programmes encouraged wetland drainage and filling to allow conversion to agriculture, settlements and industrial sites.

As a result, North America, excluding Alaska and Canada's undeveloped northern regions, lost more than one-half of its original wetland habitat with agricultural expansion responsible for between 85 and 87 per cent of the losses. Since the 1980s, wetland losses have slowed considerably. Changes in agricultural policies, particularly improvements in hydrological conditions and cooperative efforts to conserve wetlands for waterfowl were factors in these achievements.

Although more than 250 000 ha of wetlands were lost in the United States between 1986 and 1997, this was an 80 per cent reduction from the previous decade. At the global level, both countries are parties to the Ramsar Convention on Wetlands of International Importance. North America currently has 53 Wetlands of International Importance - 36 in Canada and 17 in the United States. More than 70 per cent of Canada's wetland resources are now covered by federal and provincial wetland policies and about 15 US states regulate inland wetlands.

In the United States, federal subsidies that allowed wetlands to be converted to agriculture ceased in 1985 and a Wetland Plan was issued in 1993 to make wetland regulation more fair, flexible and effective. Although past US government authority over wetlands has been fragmented and inconsistent, plans for the restoration of the Florida Everglades are testimony to the success of combined efforts among many levels of government, business and environmental NGOs.

The Canadian government does not currently track or report on the status of its wetland resources but Canada was the first nation to adopt a federal policy on wetland conservation. Wetland ecosystems make up about 17 per cent of Canada's national parks, and about 10 per cent are excluded from development. The reduction in the rate of wetland loss is a considerable achievement but wetlands are still being lost to development. The future of wetland habitat and the biodiversity it harbours may be compromised by changing conditions such as population growth, expansion of agricultural production, economic growth and changes in hydrological conditions and the flow of people.

Bio-invasion

Bio-invasion is now thought to be the second gravest threat to biodiversity in North America, after habitat destruction and degradation. Competition or predation by non-native species imperils nearly half of the species listed as threatened or endangered under the US Endangered Species Act. In Canada, alien species have been involved in causing risk to about 25 per cent of endangered, 31 per cent of threatened and 16 per cent of vulnerable species. Invasive aquatic species are particularly threatening to wetland and freshwater ecosystems and can also pose serious health risks. For example, human cholera bacteria were found in ballast tanks and in oyster and fin-fish samples in Mobile, Alabama, in 1991. Alien aquatic species are expected to contribute to the extinction of native freshwater species in North America at a rate of 4 per cent a decade over the next century.

The high economic costs of damage caused by bioinvasions in North America is causing increasing concern. Both countries have developed monitoring plans and information systems to help control bioinvasion. Responses to the challenge of invasive species include legislation, policies, and plans and programmes that focus on preventing the invasion of new species and the eradication or control of established ones. Canada and the United States cooperate in programmes related to invasive species in the Great Lakes, for example.

Despite requirements for ships to exchange ballast water at sea, however, the influx of new species into the Great Lakes continues and is considered to be a serious threat to the integrity of the Great Lakes ecosystem. As trade increases, new invasions are expected. In addition, it may be that global climate change could create conditions that are even more conducive to bioinvasion. North American as well as global cooperation is essential to stem the tide of bioinvasion and the damage it causes.

West Asian The region has wide variations in terrestrial and aquatic ecosystems. Main terrestrial habitats include Mediterranean forests, rangelands and deserts. Marine ecosystems include mudflats, mangrove swamps, sea grass and coral reefs. Rivers in the Mashriq and springs in the whole region represent freshwater ecosystems. The estimated number of endemic vascular species in the region is 800, and in some hot spots such as the Socotra

Islands of Yemen, 34 per cent of the total number of vascular plants are endemic. There are seven endemic mammal species and ten endemic birds.

The seas are rich in species diversity with 200 species of crabs, 20 species of marine mammals and more than 1,200 species of fish and more than 330 species of corals in the Red Sea and the Gulf. More than 11 per cent of the corals are endemic to the Arabian Peninsula sub-region. There are up to 12,000 marine species in the Mediterranean, representing 8-9 per cent of the world sea species richness. Substantial numbers of vertebrates are threatened with extinction in the region.

Habitat destruction and fragmentation have increased dramatically in most countries over the past three decades due to human population and resource consumption growth. Degradation of unique terrestrial and aquatic ecosystems and loss of genetic resources are the main biodiversity issues in West Asia. Water resource management and the maintenance of inland water biodiversity, as well as overhunting of large mammals and birds, are therefore among the most important issues affecting biodiversity in the region. Rapid population increases and changes in lifestyle have contributed to the degradation of wetland ecosystems due to increased exploitation of surface and groundwater.

In Jordan, groundwater extraction for urban needs increased from around 2 million m3 in 1979 to around 25 million m3 in 1993 while an additional 25 million m3 per year was used for irrigated agriculture. As well as water extraction, pollution and impacts from refugee camps in the area have led to the deterioration and drying up of the Azraq wetlands natural reserve. As a consequence tourism in Azraq has declined. In the eastern part of the

Arabian Peninsula, many of the date palm oases and natural freshwater springs have been lost in the past two decades.

By far the most serious wetland change in West Asia over the past three decades has occurred in the lower Mesopotamian marshlands, where serial satellite images confirm a loss of around 90 per cent of the area of lake and marshlands. This loss may be attributable in part to the large number of dams now present on upstream parts of the Tigris-Euphrates system, but appears to be primarily a result of major hydrological engineering works in southern Iraq, notably the completion of the Major Outfall Drain which diverts water to the head of the Gulf. However, despite some negative impacts of damming on indigenous biodiversity, the loss of some habitats such as wetlands has been offset by the creation of large artificial habitats elsewhere in the region.

Food self-sufficiency policies in the region have resulted in the cultivation of marginal lands for irrigated intensive agriculture. This has strained water resources and caused salinization, with negative effects on freshwater biodiversity. The breakdown of traditional systems of resource management has also had a major impact on biodiversity. For example, the traditional Al-Hema system, which facilitated the sustainable use of rangelands and other natural resources by setting aside large reserves during times of stress was abandoned in the 1960s in the Arabian Peninsula and Mashriq countries.

Coastal and marine biodiversity is threatened by several human activities including pollution (oil spills, industrial and domestic discharges into the sea), physical alteration of habitats (sand dredging and landfills), climate variability and alien species introduced by ballast water.

The extent of mangroves has been decreasing along the shores of the Gulf over the past 30 years due to unplanned coastal development to the extent that only 125-130 km2 of mangrove patches remain. In Saudi Arabia, more than 40 per cent of the Gulf coastline has been reclaimed and almost 50 per cent of the mangroves lost.

In the Arabian Peninsula seas, about 20,000 km2 of coral reefs or 7.9 per cent of the total area of world corals have been exposed to bleaching due to increases in sea water temperature caused by El Niño. It is feared that global warming will intensify this phenomenon. In the Mashriq sub-region many marine species, including Mediterranean monk seals, marine turtles and marine sponges, are threatened by the continuous deterioration of coastal water quality due to sedimentation, nutrient discharge and eutrophication.

A comprehensive decline in the larger terrestrial species has been recorded extinct, the Arabian bustard (Ardeotis arabs) has been reduced in numbers and is possibly extinct in Saudi Arabia, and the Houbara bustard (Chlamydotis undulata) now winters in much reduced numbers. Programmes for captive breeding of threatened species have operated since the 1980s, with re-introduction programmes for the Arabian oryx, Houbara bustard and some gazelle species in Jordan, Oman, Saudi Arabia and Syria.

Most countries have ratified the Convention on Biological Diversity. In addition, some have ratified other biodiversity-related conventions such as the Convention on International Trade in Endangered Species (CITES). Countries are also adhering to other international and regional agreements such as the Mediterranean Action Plan (MAP) and Regional Organization for the Protection of the Environment of the Red Sea and Gulf of Aden (PERSGA).

The establishment of protected areas in West Asia has been gaining momentum. Local people are generally unhappy with the existing biodiversity conservation programmes because they are not involved in decision-making. However, the situation is improving in some countries such as Lebanon and Jordan.

The polar regions face threats from climate change, ozone depletion, altered land use and the unsustainable use of natural resources. The ocean areas include some of the largest marine ecosystems on Earth and are threatened by commercial fisheries and the harvesting of marine mammals. For centuries the Arctic has attracted hunters of mammals such as whales, seals, walruses, polar bears and otters. Many species have been repeatedly driven to near extinction and some are below safe biological limits. Hunting continues but is now more tightly regulated. Even so, marine mammal populations in decline include local populations of the Beluga whale, walrus, Steller's sea lion, harbour seal, northern fur seal and the fin whale. For many more marine mammals, the trend is unknown.

Several bird populations and fish species are in decline. The latter include local populations of Atlantic cod, Arctic cod, Greenland halibut and wolf-fishes. Many wildlife populations have suffered starvation due to human activity such as overfishing. For example, in the mid-1980s the capelin stock of the Barents Sea collapsed due to overfishing, resulting in the starvation of hundreds of thousands of harp seals. At least 50 000 more were drowned in fishing gear. Norway banned capelin fishing during 1987-90, allowing the capelin population to recover and fishing to resume but at more sustainable levels.

Puffins have been another casualty. They feed their young mainly on herring fry.

In the late 1970s, some 1.4 million pairs of puffins nested at the southwestern end of the Lofoten Islands. In the 1980s, the colony contracted at a rate of 10-15 per cent a year. By 1995, it was less than half its former size because most puffin chicks starved to death due to the overfishing of herring fry since the 1960s. By the mid-1990s, puffins had still not recovered despite an increase in the herring population due to strict fishing regulations. Reducing exploitation and other responses have had positive impacts on other populations. For example, an Icelandic fishing ban on Atlantic herring between 1972-75 helped the stock to make a gradual recovery and it is now considered to be within safe biological limits.

In the 1940s, the Svalbard population of the barnacle goose had been reduced to only 300 birds. It was then totally protected on its winter ranges in the United Kingdom and a nature reserve number of known species was established. Today, there are 23 000 in the Svalbard population. Similar increases have occurred in Greenland and Russia. Other pressures on Arctic biodiversity are climate change, and habitat loss and fragmentation. The warming trend is reducing the ice habitat for species such as the polar bear and walrus, and is causing more severe climatic episodes such as ice storms that raise mortality rates. The Arctic countries have begun a major project to develop recommendations for action on the effects of global warming in the Arctic. These countries have also taken several steps to reduce habitat loss and prevent fragmentation.

The Antarctic terrestrial ecosystem is structurally simplistic with a small number of species. Marine biomass in the Southern Ocean can be immense but species richness is generally low. The benthic (bottom) fish fauna of the continental and upper slope of Antarctica includes 213 species confined to 18 families. Seals, whales and seabirds dominate the higher levels of the Southern Ocean. Knowledge of Southern Ocean marine diversity is confined largely to the continental shelves and slopes. Little is known about the fauna of the deep sea around Antarctica.

Historic sealing and whaling activities have had a significant impact on these populations in the Southern Ocean, at one point threatening extinction of some species. Today, strict international agreements govern the harvesting of Antarctic seals (Convention for the Conservation of Antarctic Seals) and whales (International Whaling Convention, which also designated large areas of the Southern Ocean as a whale sanctuary).

In 1982, the Convention on the Conservation of Antarctic Marine Living Resources (CCAMLR) was established to promote the conservation and rational use of marine living resources south of the Antarctic Convergence. Southern Ocean fisheries are now managed within the framework of CCAMLR. Although there is uncertainty in its assessments, CCAMLR estimates that the level of illegal, unregulated and unreported (IUU) fishing in the Southern Ocean -which has been a major problem for decades - had decreased in 1998 but has since increased despite stronger CCAMLR measures to combat IUU fishing.

The high level of illegal catch of Patagonian toothfish (Dissostichus eleginoides) in the South Indian Ocean has become a major concern as it threatens the sustainability of stocks. To address IUU catches, CCAMLR adopted a Catch Documentation Scheme requiring all landings, transshipments and importations of toothfish into the territories of contracting parties to be accompanied by a completed catch document. In 2000, CCAMLR took further steps to combat IUU fishing by urging all parties to avoid flagging or licensing vessels with a history of engagement in illegal practices.

CCAMLR regulation has reduced the incidental bycatch of seabirds and marine mammals in legal fisheries to low levels but illegal fishing still takes its toll. For some seabird populations, long-line fisheries represent a major threat. This led to the listing in 1997 of all albatross species on the protected species list of the CMS. Albatross and petrel species, such as the wandering albatross and the Antarctic giant petrel, have also been listed as vulnerable in the IUCN Red List. The final draft of the Agreement on the Conservation of Albatross and Petrels was recently completed in Cape Town, South Africa.

Changes to the distribution and composition of terrestrial flora and fauna, attributable to recent warming over Antarctica, have been observed over the past three decades. Composition and distribution of marine species is also expected to change with a changing climate. It has been indicated that the marked increase in the number of Adelie penguins in the Ross Sea area in the 1980s showed remarkable synchronization with the climatic variation in the same region. At Palmer Station on Anvers Island, where Adelie penguins are known to have nested only prior to the 1950s, gentoo and chinstrap penguins are now breeding and have expanded their ranges southward in the Peninsula within the past 50 years, in correlation with pronounced regional warming.

Changes in the extent and thickness of ice affect the timing, magnitude and duration of the seasonal pulse of primary production in the polar regions. It has been suggested that sea ice extent affects krill availability which in turn may affect krill predators. Regional warming and reduced krill abundance may therefore affect the marine food web. The density and abundance of minke whales has been observed to be lower in seasons with warmer sea surface temperatures, fewer cold-water intrusions, and smaller sea ice extent, possibly owing to the shift in availability of prey. Along the west coast of the Antarctic Peninsula, springtime ozone depletion can lead to a twofold increase in biologically effective UV-B radiation.

UV exposure affects phytoplankton, including inhibition of primary production. This is a major concern in view of the phytoplankton's key role in the short food chain of the Antarctic marine ecosystem. The spring bloom of phytoplankton coincides with the springtime ozone hole and the subsequent period of high UV-B radiation. Reduction of phytoplanktonic production associated with the ozone hole is estimated to be 6-12 per cent.

7

Atmosphere

The World Health Organization (WHO) lists six 'classic' air pollutants: CO, lead, nitrogen dioxide (NO_2), suspended particulate matter (SPM) -including dust, fumes, mists and smoke - SO_2 and tropospheric ozone (O_3). The burning of fossil fuels and biomass is the most significant source of air pollutants such as SO_2, CO, certain nitrous oxides such as NO and NO_2 (known collectively as NOx), SPM, volatile organic compounds (VOCs) and some heavy metals. It is also the major anthropogenic source of carbon dioxide (CO_2), one of the important greenhouse gases.

Between 1973 and 1998, total energy supply increased by 57 per cent, the majority provided by oil, natural gas and coal with nuclear and hydropower and other renewable resources playing a minor role. The fuels used vary from region to region -for example, natural gas dominates in the Russian Federation, while coal provides 73 per cent of the energy consumed in China. Biomass is an important source of energy in the developing world and is the main source of indoor air pollution in such countries.

Acid precipitation has been one of the most important environmental concerns over the past decades especially in Europe and North America, and more recently also in China. Significant damage to forests in Europe became a high priority environmental issue around 1980, while thousands of lakes in Scandinavia lost fish populations due to acidification from the 1950s to the 1980s. In some parts of Europe, the anthropogenic SO_2 emissions which lead to acid precipitation have been reduced by nearly 70 per cent from their maximum values; there have also been reductions of some 40 per cent in the United States. This has resulted in a significant recovery of the natural acid balance, at least in Europe. On the contrary, as a result of the growing use of coal and other high sulphur fuels, increasing SO_2 emissions in the Asia and Pacific Region are a serious environmental threat.

Air pollutant emissions have declined or stabilized in most industrialized countries, largely as a result of abatement policies developed and implemented since the 1970s. Initially governments tried to apply direct control instruments but these were not always costeffective. In the 1980s, policies were directed more towards cost-effective pollution abatement mechanisms that relied on a compromise between the cost of environmental protection measures and economic growth. The Polluter Pays Principle has become a basic concept in environmental policy planning.

Recent policy developments, at both national and regional levels, are based on economic and regulatory instruments, and technology improvement and transfer to enhance emission reductions. In the international arena, one of the most important political developments has been the Convention on Long-Range Transboundary Air

Pollution (CLRTAP), adopted in 1979. Through a series of protocols establishing reduction objectives for the main air pollutants, this treaty has catalysed European, Canadian and US governments to implement national emission abatement policies.

The most recent protocol is the 1999 Protocol to Abate Acidification, Eutrophication and Ground-Level Ozone that sets up new reduction commitments for SO_2, NOx, VOCs and ammonia (NH_3). Stricter environmental regulation in the industrialized countries has triggered the introduction of cleaner technology and technological improvements, especially in the power generation and transport sectors.

In the latter, a significant reduction in harmful emissions has been achieved due to the improved engine combustion cycle, increased fuel efficiency and the widespread introduction of catalytic converters. In developing countries, however, the sources of emissions are more varied and include highly polluting power plants, heavy industry, vehicles and the domestic combustion of coal, charcoal and biomass. While the emission of pollutants can be significantly reduced for a small cost, few developing nations have made even small investments in pollution reduction measures, even though the environmental and population health benefits of such measures are evident.

Though measurable progress in industrial emission abatement has been achieved at least by the developed countries, transport has become one of the major sources of air pollution (particularly NO_x and many carbon compounds) in many countries. High concentrations of these compounds in urban air can, under certain climatic conditions, result in photochemical smog that severely affects human health. In many urban centres and their surrounding areas, high concentrations of tropospheric O_3 are an additional problem.

Anthropogenic tropospheric ozone may be produced by reactions between NOx and VOCs on warm sunny days, especially in urban and industrial areas and in regions prone to stagnant air masses. This can have far-reaching implications since molecules of O_3 have been found to travel large distances from emission sources. Tropospheric O_3 concentrations over large areas of Europe and some areas of North America are so high that not only is human health threatened but vegetation is also affected.

For example, in the United States ground-level ozone has been estimated to cost more than US$500 million a year as a result of reductions in agricultural and commercial forest yields. Urban air pollution is one of the most important environmental problems. In most European and North American cities, the concentrations of SO_2 and SPM have decreased substantially in recent years.

However, in many developing countries, rapid urbanization has resulted in increasing air pollution in many cities, WHO air quality guidelines are often not met and, in megacities such as Beijing, Calcutta, Mexico City and Rio de Janeiro, high levels of SPM prevail. A final issue of global concern is that of persistent organic pollutants (POPs). These substances are known to decay slowly and they can be transported over long distances through the atmosphere. High concentrations of some POPs are found in polar areas with potentially serious regional environmental impacts. These compounds can also accumulate in animal fats, representing a health risk.

The Stockholm Convention on Persistent Organic Pollutants, adopted in May 2001, sets out control measures covering handling of pesticides, industrial chemicals and unintended byproducts. The control provisions call for eliminating the production and use of intentionally produced POPs, and eliminating unintentionally produced POPs where feasible.

Stratospheric Ozone Depletion

The protection of the Earth's ozone layer has presented one of the major challenges over the past 30 years, spanning the fields of environment, trade, international cooperation and sustainable development. The thinning of the ozone layer threatens human health through diseases such as skin cancer, eye cataracts and immune deficiency, affects flora and fauna, and also influences the planet's climate. Ozone depletion is brought about by a number of chemicals known as ozone-depleting substances (ODS), the most notorious of which are the chlorofluorocarbons (CFCs).

In 1974, the results of studies linking stratospheric ozone depletion to the release of chloride ions in the stratosphere from CFCs were made publicly available. ODS are used in refrigerators, air conditioners, aerosol spray, insulating and furniture foams, and firefighting equipment, and their production peaked in the late 1980s as the demand for such goods grew. The depletion of the Earth's ozone layer has now reached record levels, especially in the Antarctic and recently also in the Arctic.

In September 2000, the Antarctic ozone hole covered more than 28 million square kilometres. Current average ozone losses are 6 per cent in the northern mid-latitudes in winter and spring, 5 per cent in southern mid-latitudes all year round, 50 per cent in the Antarctic spring and 15 per cent in the Arctic spring. The resulting increases in harmful ultraviolet irradiation amount to 7 per cent, 6 per cent, 130 per cent and 22 per cent respectively. However, due to continuous efforts by the international community, the global consumption of ODS has decreased markedly and the ozone layer is predicted to start recovering in the next one or two decades and to return to a pre-1980 level by the middle of the 21st century if all the future control measures of the Montreal Protocol are adhered to by all countries.

International cooperation has been the key to protecting the stratospheric ozone layer. Nations agreed in principle to tackle a global problem before its effects became evident or its existence scientifically proven - probably the first example of acceptance of the precautionary approach. International action began in earnest in 1975 when the UNEP Governing Council called for a meeting to coordinate activities on protecting the ozone layer. A Coordinating Committee on the Ozone Layer was established the following year to undertake an annual scientific review.

In 1977, the United States banned use of CFCs in non-essential aerosols. Canada, Norway and Sweden soon enacted similar control measures. The European Community (EC) froze production capacity and began to limit use of aerosols. These initiatives, though useful, provided only a temporary respite. After falling for several years, CFC consumption began increasing again in the 1980s, as non-aerosol uses, such as foam blowing, solvents and refrigeration, increased. Stricter control measures were needed and UNEP and several developed countries took the lead, calling for a global treaty on stratospheric ozone layer protection.

The Vienna Convention for the Protection of the Ozone Layer was finally agreed by 28 countries in March 1985. It encouraged international cooperation on research, systematic observation of the ozone layer, monitoring of ODS production, and the exchange of information. In September 1987, 46 countries adopted the Montreal Protocol on Substances that Deplete the Ozone Layer. The original Protocol required only a 50 per cent cut in consumption of five widely used CFCs by December 1999, and a freeze in the consumption of three halons.

Most ODS - including all the substances specified in the original Protocol - were phased out in industrialized countries by the end of 1995. The Protocol provides a 10-year grace period for developing countries and the financial mechanism to meet the costs to these countries of phasing out ODS, thus realizing the principle of common but differentiated responsibility. By 2000, the Multilateral Fund had disbursed more than US$1.1 billion for capacity building and projects to phase out ODS in 114 developing countries. Almost every party to the Montreal Protocol has now taken measures to phase out ODS with the result that, by 2000, the total consumption of ODS had been reduced by 85 per cent.

Greenhouse Gases and Climate Change

Scientists have known about the natural 'greenhouse effect' for more than a century: the Earth maintains its equilibrium temperature through a delicate balance between the incoming solar energy it absorbs and the outgoing infra-red energy (long wavelength radiation) that it emits and some of which escapes into space. Greenhouse gases (water vapour, carbon dioxide, methane and others) allow solar radiation to pass through the Earth's atmosphere almost unimpeded but they absorb the infra-red radiation from the Earth's surface and then re-radiate some of it back to the Earth.

This natural greenhouse effect keeps the surface temperature about 33°C warmer than it would otherwise be - warm enough to sustain life. Since the industrial revolution, the concentration of CO_2, one of the major greenhouse gases, in the atmosphere has increased significantly. This has contributed to the enhanced greenhouse effect known as 'global warming'. The CO_2 concentration in the atmosphere is currently about 370 parts per million (ppm) - an increase of more than 30 per cent since 1750. The increase is largely due to anthropogenic emissions of CO_2 from fossil fuel combustion and to a lesser extent land-use change, cement production and biomass combustion.

Although CO_2 accounts for more than 60 per cent of the additional greenhouse effect accumulated since industrialization, the concentrations of other greenhouse gases such as methane (CH_4), nitrous oxide (N_2O), halocarbons and halons have also increased. In comparison to CO_2, CH_4 and N_2O have contributed about 20 per cent and 6-7 per cent respectively to the additional greenhouse effect. Halocarbons have contributed about 14 per cent. Many of these chemicals are regulated under the Montreal Protocol.

However, those which have negligible ozone-depleting potential are not controlled under the Montreal Protocol. Although they have accounted for less than 1 per cent of the additional greenhouse effect since industrialization, their concentrations in the atmosphere are increasing. Greenhouse gas emissions are unevenly distributed between countries and regions. In general, industrialized countries are responsible for the majority of historical and current emissions.

OECD countries contributed more than half of CO_2 emissions in 1998, with an average per capita emission of about three times the world average. However the OECD's share of global CO_2 emissions has decreased by 11 per cent since 1973. In assessing the possible impact of rising atmospheric concentrations of greenhouse gases, IPCC concluded in 2001 that 'there is new and stronger evidence that most of the warming observed over the last 50 years is attributable to human activities'.

Ecosystems, human health and economy are all sensitive to changes in climate - including both the magnitude and rate of climate change. Whereas many regions are likely to experience adverse effects of climate change—some of which are potentially irreversible - some effects could be beneficial for some regions. Climate change represents an important additional stress on those ecosystems already affected by increasing resource demands, unsustainable management practices and pollution. Some of the first results of the changing climate can serve as indicators.

Several vulnerable ecosystems such as coral reefs are seriously endangered by increased sea temperature and some populations of migratory birds have been declining because of unfavourable variations in climatic conditions. Climate change is furthermore likely to affect human health and well-being through a variety of mechanisms. For example, it can adversely affect the availability of freshwater, food production, and the distribution and seasonal transmission of vector-borne infectious diseases such as malaria, dengue fever and schistosomiasis.

The additional stress of climate change will interact in different ways across regions. It can be expected to reduce the ability of some environmental systems to provide, on a sustained basis, key goods and services needed for successful economic and social development, including adequate food, clean air and water, energy, safe shelter and low levels of diseases. The United Nations Framework Convention on Climate Change (UNFCCC) adopted at UNCED in 1992 has the ultimate objective of 'stabilization of greenhouse gas concentrations in the atmosphere at a level that would prevent dangerous anthropogenic interference with the climate system'.

The Convention further defines several principles of fundamental importance, for example that parties should take precautionary measures and act 'on the basis of equity and in accordance with their common but differentiated responsibilities'. Being a framework treaty, the UNFCCC contained only a non-binding recommendation for industrialized countries to return to the 1990 emission levels of CO_2 and other greenhouse gases by the year 2000. However, most of them have not returned anthropogenic emissions of greenhouse gases to 1990 levels.

In general, global emissions of almost all anthropogenic greenhouse gases, particularly CO_2, continue to increase. This reflects the inadequacy of national and international policies and measures to address climate change. Overall, industrialized countries are required to reduce their aggregated emissions to at least 5 per cent below the 1990 level in the period 2008-12. No new obligations were introduced for developing countries. The Kyoto Protocol also allows collective implementation of obligations by means of applying the so-called 'Kyoto mechanisms'. These mechanisms aim at providing 'geographical flexibility' and reducing the costs of complying with the Kyoto targets.

In view of anticipated economic losses, some industrialized nations have prejudiced the Kyoto commitments and the Kyoto Protocol as a whole. The Kyoto Protocol would never have come into force if other developed countries had adopted the same position. However, at the resumed 6th Conference of the Parties in Bonn, Germany, in July 2001, the parties successfully completed negotiations aimed at setting the operational details for commitments on reducing emissions of greenhouse gases. They also reached agreement on actions to strengthen implementation of the UNFCCC itself. The political decision - or Bonn Agreement - was formally adopted by the COP on 25 July 2001. Many saw it as an 'historic' political agreement that saved the Kyoto Protocol and paved the way to its ratification, though it was clearly recognized that this was just a small step towards solving the global problem.

The African continent is climatically diverse. Humid tropical conditions prevail in Western and Central Africa and in the Western Indian Ocean islands; most Southern African countries experience arid and semiarid conditions, while semi-deserts and deserts are features of Northern Africa. The region experiences a high degree of variability and uncertainty in climatic conditions. Climate variability is therefore the single most important atmospheric phenomenon in Africa.

In comparison to other regions, African countries emit negligible amounts of air pollutants and anthropogenic greenhouse gases. For example, Africa contributes less than

3.5 per cent of global emissions of CO_2. Nevertheless, anthropogenic atmospheric pollution is a problem in Northern and Southern Africa, and in some large cities. South Africa contributed 42 per cent of the total regional emissions of CO_2 in 1998; some Northern African countries, where total energy consumption increased by 44 per cent from 1980 to 1998, also contribute significantly.

Subsidizing electricity production, promoting industrial development economic strategies and increased consumption have contributed to rising emissions in some areas. For example, in Mauritius total energy consumption doubled between 1990 and 1998, and CO_2 emissions rose 23 per cent from 1991 to 1995. A rapid increase in the number of private cars, and the poor condition of many commercial and private vehicles, are additional causes of concern. Vehicle emissions are the major source of lead contamination, and contribute to dust, noise and smoke pollution.

Policies are in place in Algeria, Mauritius and Morocco to encourage conversion to newer, less polluting vehicles, and unleaded petrol has been promoted or even subsidized in Egypt, South Africa and Tunisia. Industrial processes are also significant sources of atmospheric pollution, especially in large urban centres where the pollutants sometimes combine to create atmospheric smog. In Northern Africa, the Western Indian Ocean islands, Southern Africa and some large cities, the incidence of respiratory disease is increasing, reflecting a deterioration of air quality.

In sub-Saharan Africa, traditional fuels accounted for 63.5 per cent of total energy use in 1997. Many countries have prepared National Environmental Action Plans (NEAPs) or National Strategies for Sustainable

Development (NSSDs) which address, among other things, sources and impacts of atmospheric pollution. Ghana, Kenya, South Africa, Uganda and Zambia are among those that have introduced legislation making Environmental Impact Assessments (EIAs) compulsory for developments such as roads, mines and industrial operations with potentially high atmospheric emissions.

Climatic variability

Climatic variability and associated floods and droughts result in increased risks of crop failure and therefore reduced food security, as well as higher incidences of malnutrition and disease. In Ethiopia, for example, the 1984 drought affected 8.7 million people, 1 million people died and millions more suffered from malnutrition and famine. This drought also caused the death of nearly 1.5 million livestock. The 1991-92 drought in Southern Africa caused a 54 per cent reduction in cereal harvest and exposed more than 17 million people to the risk of starvation. More than 100 000 people died in the Sahelian drought of the 1970s and 1980s.

Crop failure and livestock losses lead to increased dependence on imports and foreign aid, reducing economic performance and the ability to cope with future environmental disasters. In 1997 and 1998, parts of Eastern Africa suffered from high rainfall and flooding due to ENSO disturbances, and in 1999 and 2000 Southern Africa and the Western Indian Ocean islands experienced devastating cyclones and floods. Flood water is an ideal habitat for bacteria and mosquitoes. In Uganda, the ENSO-induced floods of 1997-98 caused more than 500 deaths from cholera, and a further 11 000 people were hospitalized.

According to IPCC, Africa is the most vulnerable region in terms of predicted decreases in water and food security, because widespread poverty limits adaptive capacity. Changes in rainfall could also have serious consequences for those parts of Africa that depend on hydroelectricity. The anticipated sea level rise resulting from global climate change may threaten many coastal settlements and islands including the Western Indian Ocean islands.

Even if anthropogenic greenhouse gas emissions were stabilized immediately, sea level would continue to rise for many years. IPCC also predicts that the intensity of cyclones, rain and wind will probably increase, and the cyclone zone in the Western Indian Ocean could expand to include Seychelles. Changes to rainfall and temperature patterns could also alter biodiversity, with many species not being able to adapt or migrate to more suitable areas. WWF forecasts that an anticipated 5 per cent decrease in rainfall in Southern Africa will affect grazing species such as hartebeest, wildebeest and zebra, threatening wildlife in the Kruger National Park, South Africa, the Okavango delta in Botswana and Hwange National Park in Zimbabwe.

There are also fears that malaria could spread to new areas such as parts of eastern Namibia and northern South Africa. The region's ability to adapt to climate change will depend on several factors, including population growth and consumption patterns, which will affect demand for food and water, and the location of populations and infrastructure in relation to vulnerable coastal areas, which will determine economic losses due to sea level rise. Many countries will need to change their agricultural practices, particularly to reduce dependency on rainfed agriculture, and to avoid cultivation in marginal areas.

Rural communities that currently depend on biomass for energy may be forced to seek alternative sources if climate change brings about changes to vegetation type and distribution. Almost all African countries have ratified the UNFCCC and many are in favour of the Kyoto Protocol. African countries stand to benefit from mechanisms of international cooperation proposed under the Protocol. Countries that are rich in natural forests may also enter into emissions reduction transfer agreements with industrialized countries, contributing to their own economies and assisting development.

On the whole, African countries have an interest in a decision on mechanisms that ensures they facilitate sustainable development in Africa, produce gains in the climate system, contribute towards Africa's adaptation to climate change and result in projects that accelerate socio-economic growth. South Africa emits the most carbon in Africa but, being classified as a developing country, is not formally required to control emissions of greenhouse gases. However, a National Committee on Climate Change has been established to oversee research, communication and the development of policy on climate change.

In both Northern and Southern Africa, options for further exploitation of alternative sources of energy are being explored. Such initiatives are likely to be most successful in remote areas, where connections to centralized power sources are expensive, and electricity demand is for domestic use only. One of the main challenges for most African countries with regard to climatic and atmospheric changes is the need to adapt development processes to the changes in the environment.

Mitigation and adaptation mechanisms need to be developed to cope with the impacts of changes in weather patterns and intensified droughts and floods associated with El Niño events. On the other hand, African countries may contribute to the implementation of the UNFCCC and Kyoto Protocol by adopting energy efficient and renewable energy technologies. The Kyoto mechanisms and international institutions to be created to realize these mechanisms provide for active participation of African countries. Failure to realize the potentialities of the Kyoto Protocol may result in further exposure to the adverse effects of climate change.

Localized indoor and outdoor pollution must be controlled and abated through the introduction of cleaner industrial processes, improved transport systems and waste management. For example, a waste management strategy that envisages reducing illegal dumping and burning of solid waste has been developed in South Africa.

A serious environmental issue in the Asia and Pacific region is the rapid degradation of the quality of air. Depletion of the ozone layer and the consequences of global climate change are also serious. Air pollution levels in the most populated cities are amongst the highest in the world, producing serious human health impacts and affecting aquatic and terrestrial ecosystems. Transport is a significant, often major, source of urban air pollution. The other sources include industrial emissions, burning of solid and liquid fuels for power generation, and burning of biomass and other fuels such as charcoal for household use.

In a few cities, pollution levels have decreased. For example, in Japan, high fuel prices, technological advances and strict standards have reduced SO_2 and particulate emissions, and eliminated lead emissions from transport. However, NO_x emissions in Tokyo and Osaka have not declined sufficiently because of an increasing number of vehicles. This situation is common in cities with growing levels of private transport. Traffic has become a major air polluter in the big cities, although most Asian countries have low per capita vehicle ownership in comparison to the world average.

Many countries have developed their own air quality standards for principal pollutants as well as emission standards for power plants, selected industries and vehicles. To reduce pollution, many countries have introduced unleaded petrol, mandatory catalytic converters and low sulphur motor fuels. Alternative technologies such as electric vehicles and compressed natural gas operated vehicles are also being considered, especially in India and the Islamic Republic of Iran. Tax incentives for gas or battery operated vehicles have been introduced in Nepal and Pakistan.

Except for developed countries of the Northeast Asian and parts of the South Pacific sub-regions, concern over the environment began to create a demand for improved environmental protection in the 1980s. The 1990s have seen significant progress in establishing the institutions and policy tools needed to address urgent environmental problems. However, continuous growth in energy consumption and reliance on fuels such as coal and oil with a relatively high carbon content will inevitably increase emissions unless more aggressive policies are introduced.

Indoor air pollution is often a more severe health hazard than outdoor air pollution. Most rural inhabitants in the region use twigs, grass, dried animal dung, crop residues, wood, charcoal and kerosene as household fuels. Coupled with inadequate ventilation, this results in highly contaminated indoor air. Given the high levels of harmful emissions and the number of people using traditional cooking fuels - Asia produces nearly half of the world's woodfuel - the scale of exposure is large.

Health effects include acute respiratory infection in children, chronic obstructive lung disease, adverse pregnancy outcomes and lung cancer in women. Acute respiratory diseases are prevalent in the rural and/or hilly areas of Afghanistan, Bangladesh, Bhutan, India, Nepal, Pakistan and Sri Lanka where indoor air pollution is high. About 40 per cent of the global infantile mortality caused by pneumonia occurs in Bangladesh, India, Indonesia and Nepal; many of these deaths are caused by pollutants from burning traditional fuels.

In India, household solid fuel use is estimated to cause about 500 000 premature deaths a year in women and children under five. There are indications that tuberculosis and blindness may be associated with indoor air pollution. Indoor air pollution is blamed for 5-6 per cent of the national burden of diseases in women and children in India. The key areas for intervention are: use of cleaner fuels such as low-propane gas and kerosene; development of high-grade biomass fuels; improvements in stove design and better dissemination of stoves; improvements in housing; and improvements in environmental awareness and education.

To address the problem of indoor air pollution in India, around 3 million biogas plants and more than 22 million improved cooking stoves have been installed in rural and remote areas of the country, resulting in a saving of the equivalent of 21 million tonnes of firewood per annum. Haze and acid rain have been the emerging regional issues over the past decade, especially in Asia because of China and India's overwhelming dependence on coal. Around 0.28 million ha of forest land are reported to be damaged by acid rain in the Sichuan basin of China. SO2 emissions in Asia are estimated to have increased from about 26.6 to about 39.2 million tonnes during 1985-1997.

In China, a reduction of 3.7 million tonnes or 15.8 per cent in SO_2 emissions was achieved during 1995-2000. At least two-thirds of acid depositions in the region are caused by coal-fired power plants with outdated pollution control equipment. Haze problems are also prevalent in the region due to forest fires in Southeast Asia. The most serious episode occurred in 1997, when the effects of forest fires in Indonesia extended to neighbouring countries including Brunei Darussalam, Papua New Guinea, Philippines, Singapore and Thailand.

Ozone depletion

Depletion of the stratospheric ozone layer has emerged as a serious concern in the region. Data from Australia and New Zealand show that ultraviolet levels there appear to be rising by about 10 per cent per decade. It follows that the average exposure time for an individual in Australia to develop sunburn has been reduced by approximately 20 per cent over the past 20 years. India and China are the largest regional producers and users of CFCs. China's consumption of ODS increased more than 12 per cent per year during 1986-94. India is the second largest producer and the fourth largest consumer of CFCs in the world.

The Multilateral Fund of the Montreal Protocol and GEF have been helping the region meet the goals of the Montreal Protocol. China has made a commitment to phase out the consumption of ODS by 2010. It has already banned the establishment of new CFC- and halon-related production facilities, and developed general and sector-specific plans with the help of the World Bank and the Multilateral Fund. The latter has approved a World Bank project which will help India phase out CFC production by 2010.

The countries of Central Asia have also made considerable progress. Azerbaijan, Turkmenistan and Uzbekistan are working towards the phase out of the use of ODS during 2001-03. Per capita use of commercial energy increased annually by 1.9 per cent in South Asia and 3 per cent in East Asia and the Pacific during 1980-98. CO_2 is the main anthropogenic greenhouse gas. Methane emissions are also high in South Asia, accounting for approximately 50 per cent of the total global anthropogenic emission of CH_4.

In New Zealand, CH_4 emissions are an order of magnitude higher than the global per capita average, primarily due to the large number of ruminant animals. Areas under most threat from climate change include marine ecosystems, coastal systems, human settlements and infrastructure. Countries in the Northwest Pacific and East Asia sub-regions and the Pacific Island countries may be particularly vulnerable to phenomena such as sea level rise because many of their human settlements and industrial facilities are located in coastal or lowland areas.

For the small island developing states, climate change and extreme weather events may have dramatic impacts on terrestrial biodiversity, subsistence cropping and forest food sources. The densely populated and intensively used low-lying coastal plains, islands and deltas in South Asia are especially vulnerable to coastal erosion and land loss because of inundation and sea flooding, upstream movement of the saline/freshwater front and sea water intrusion into freshwater aquifers.

GEF and UNDP are promoting projects to help countries in the region assess their emissions and formulate strategies to reduce them. For example, countries participating in the Asia Least Cost Greenhouse Gas Abatement Strategies project have identified a number of options in the energy sector to reduce greenhouse gas emissions.

Air pollution was one of the threats to human health and ecosystems that was recognized early in Europe. A treaty was signed as early as the late 1970s and entered into force in 1983 to curb anthropogenic emissions of harmful substances. The main sectors and activities driving air pollution in Western Europe in the past three decades have been energy, transport, industry, agriculture, solvent use, and storage and distribution of fossil fuels.

In Central and Eastern European (CEE) countries, the power and heavy industry sectors have traditionally been the major polluters, with transport only significant in major cities. In the early 1990s, economic recession was a driver in the decrease of air pollution in CEE but at the same time there was a sharp growth in the use of private cars. For example, even during the worst recession years, the number of private cars in Armenia, Russia and Ukraine increased by more than 100 per cent. This rapid increase in private car ownership has made transport an increasingly important contributor to CEE's air quality problems.

Emissions of most key air pollutants have declined over the whole of Europe since the early 1980s. By the end of 2000, emissions of sulphur compounds had been reduced to less than one-third of 1980 levels in Western Europe, and to two-thirds of those levels in CEE. A significant recovery of natural acid balance of water and soils has been observed in Europe, mainly due to reductions in SO_2 emissions, although the emissions are still too high to avoid serious effects in sensitive ecosystems. Average figures, however, mask a wide variation among countries and sub-regions.

For instance, SO_2 emissions increased by 7 per cent in Greece and 3 per cent in Portugal between 1990 and 1998 while reductions of 71 per cent and 60 per cent were observed in Germany and Finland respectively. NOx and NH3 emissions have not decreased significantly in Western Europe except for NO_x in Germany and the United Kingdom but NOx has been reduced in many CEE countries.

A lack of monitoring of emissions of heavy metals, POPs and SPM, especially in CEE countries, means that no convincing trends can be observed but it is clear that particulate matter and tropospheric ozone precursors still represent serious problems. In Western Europe, emissions of SO_2, NO_x and NH_3 have shown a clear de-coupling from GDP growth, pointing towards some degree of effectiveness of measures taken.

In some of the CEE countries that are likely to be in the first wave of accession to the European Union (EU), economic restructuring and environmental actions also appear to have had an effect in reducing air pollution. In other CEE countries, the fall in industrial output due to the recession appears to have been the main factor in air pollution reduction.

In countries such as Russia and Ukraine, emissions per unit of GDP have actually increased but the effect has been masked by the overall fall in GDP. It is clear that the reductions in emissions are at least partly due to national and local measures that have been taken to achieve targets set by CLRTAP and its Protocols, and to EU Directives linked to air emissions such as the Limitation of Emissions of Certain Pollutants into the Air from Large Combustion Plants Directive and various directives on vehicle emissions, the change to unleaded petrol and higher quality diesel fuels and improved engine design. Despite this clear progress, many air pollution reduction targets have still not been met.

In Western Europe, only the EU and CLRTAP targets for SO_2 were met well before the target date with less progress on NO_x, NH_3 and VOCs. Two recent European measures are expected to achieve further reductions in air pollutants: a proposal for an EU Directive on National Emission Ceilings for Certain Atmospheric Pollutants (NECD) and the CLRTAP Protocol to Abate Acidification, Eutrophication and Ground-level Ozone.

In many European countries, additional measures will be required to achieve the NECD and CLRTAP Protocol targets. In Western Europe 'non-technical' measures for controlling pollution such as road pricing and tax incentives have become more important but in many CEE countries it is unlikely that currently weak environmental protection bodies can enforce an effective air pollution reduction strategy in the near future.

The thickness of the ozone layer over Europe has decreased measurably since the 1980s. Though the average ozone loss in northern mid-latitudes is 6 per cent in winter and spring, at certain times the loss can be more significant. For example, in the spring of 1995 after an unusually cold Arctic winter, stratospheric ozone concentrations over Europe were 10-12 per cent lower than in the mid-1970s. The winter of 1995-96 was even colder and ozone concentrations over the United Kingdom fell by almost 50 per cent in the first week in March, the lowest ever recorded there.

As a result, an increase in ultraviolet radiation occurred in Europe between 1980 and 1997, with a clearly higher increase in the northeast. As a result of implementing the Vienna Convention and its Montreal Protocol, ODS production in Western Europe has decreased by almost 90 per cent while the production of hydrochlorofluorocarbons (HCFCs) has increased. The political and economic transition in CEE has delayed the phase out of ODS production and consumption but there is progress.

Large amounts of donor assistance were channelled via the GEF in the 1990s to upgrade technologies in favour of ozone-friendly substances. An important milestone for the Countries with Economies in Transition (CEIT) was the ceasing of production of ODS listed in Annexes A and B to the Montreal Protocol by the Russian Federation, the major regional producer, in December 2000. Although many European countries are enthusiastic proponents of a global climate change treaty, the region is still a major emitter of anthropogenic greenhouse gases.

.Most CO_2 emissions come from fossil fuel combustion. The energy sector (power and heating) is the main contributor, while transport, combustion, manufacturing and heavy industry also play a major role. Greenhouse gas emissions in the EU decreased by 2 per cent between 1990 and 1998, mainly as a result of stabilization of CO_2 emissions and reduction of emissions of N_2O and methane. Most of this decrease was attributable to Germany (as a result of increased efficiency of new power plants, energy saving in households and industries, and economic restructuring in the former East Germany) and the United Kingdom (following a switch from coal to gas).

In Western Europe generally, there has been a clear de-coupling between emissions, economic growth and energy consumption thanks to a combination of increases in energy efficiency, and effects of policies and measures to reduce greenhouse gas emissions. However, meeting the Kyoto Protocol targets will still be difficult.

Economic transformation in CEE has contributed to a significant decrease of anthropogenic greenhouse gas emissions. In 2000, CO_2 emissions in nine of those countries were 8 per cent lower than in 1990. In some of the CEE countries, economic restructuring and environmental actions appear to have had an effect in reducing CO_2 while in most CEE countries recession and a decline in industrial output appears to have been the main factor in CO_2 reduction.

Air pollution is one of the most critical environmental issues in the Latin America and Caribbean Region because of its impact on human health, especially in urban areas. Rapid urbanization, population growth, industrialization and a growing number of motor vehicles are the main causes of air pollution. The region is also prone to the negative impacts of stratospheric ozone layer depletion.

Some three-quarters of the population of Latin America and the Caribbean live in cities. Several megacities such as Buenos Aires, Mexico City, Rio de Janeiro and São Paulo, each with a population of more than 10 million, are located in the region and economic growth in these urban centres has caused increases in air pollution and associated human health impacts.

Today the problem extends beyond large cities and also affects medium-size cities and small islands. The transport sector is a major source of urban air pollution - 70 per cent of emissions in Buenos Aires and Mexico City are transport-related, with the number of cars in Mexico City increasing fourfold between 1970 and 1996. Industry, agriculture and the municipal sector also contribute to air pollution. In Santiago, the most significant sources of air pollution are transport, and small and mediumsize enterprises.

In addition, unfavourable topographic and meteorological conditions in some cities aggravate the impact of pollution: the Valley of Mexico obstructs the dispersal of pollutants from its metropolitan area as do the hills surrounding Santiago. The growth of industry, agriculture and transport sectors over the past 30 years has been accompanied by a steady increase in CO_2 emissions - estimated at 65 per cent between 1980 and 1998.

In 1991-92, the region was estimated to be responsible for some 11 per cent of global anthropogenic emissions of CO_2 - 4.5 per cent of global industrial emissions and 48.5 per cent of emissions from landuse change. Deforestation is thought to be the principal cause of emissions in the region, particularly in the Amazon basin. Deforestation and livestock breeding also result in a huge regional methane emission - some 9.3 per cent of the world total. The average annual per capita carbon emission from industry in the region was 0.73 tonnes in 1998, somewhat lower than the world average of 1.06 tonnes.

Mexico is the region's largest carbon emitter. Industrial pollutants originate mostly from fuel combustion processes in the power generation sector, although emissions of heavy metals such as lead and mercury are also important. In oilproducing countries, emissions from the refining process are also significant - for example, in Mexico City, almost 60 per cent of SO_2 emissions originate from industry, including oil refineries in the metropolitan area.

In many countries, mining activities result in local deterioration of air quality. Other sources of air pollution have local and subregional impacts, including pesticide use in agriculture and airborne particles resulting from soil erosion and biomass combustion. Studies conducted in Colombia and Ecuador in the early 1990s revealed that more than 60 per cent of agricultural workers involved in production for foreign markets had symptoms of acute pesticide poisoning (headaches, allergies, dizziness, dermatitis, blurred vision) while others experienced serious chronic effects (stillbirths, miscarriages, and respiratory and neurological problems).

Neighbouring inhabitants can also be affected, as has been found for Nicaraguan cotton fields and Costa Rican coffee plantations. Forest fires are another important contributor to air pollution, sometimes having a significant longdistance effect. In Latin America and the Caribbean, about onefifth of the population uses biomass as a major household fuel, resulting in indoor air pollution. This mainly affects women, children and the elderly who stay indoors for long periods. In Colombia and Mexico, for example, women using biomass for cooking are up to 75 times more likely to contract chronic lung disease than the average person.

Air pollution is blamed for 2.3 million annual cases of infantile chronic respiratory sickness and 100 000 cases of chronic adult bronchitis in the region. In recent decades, significant efforts have been made to cope with air pollution, especially in urban areas through strategies that include emission controls, changes in fuels and contingency controls. In Santiago, SPM emissions - and the number of days when alerts were sounded or emergencies declared - have been reduced considerably over the past decade, with concentrations of PM10 and PM2.5 decreasing by 24.1 per cent and 47.4 per cent respectively between 1989 and 1999.

This is largely due to implementation of a plan initiated in 1990 which included control of household and industrial emissions, developing monitoring capacity, removing highly polluting buses, controlling bus circulation and emissions, introducing car catalytic converters, improving motor fuel quality, and paving streets. Notwithstanding the progress achieved, urban air pollution is a serious concern even in some small and medium-sized cities because of the continued growth of the transport and industrial sectors, coupled with a lack of adequate monitoring and regulations.

The growth of vehicles due to rising real incomes and the removal of tariffs could nullify the progress made in improving air quality. By 2010, 85 per cent of the population are expected to be living in urban areas, and combating air pollution and preventing its negative health impacts will be a priority in every country. Ozone depletion is a significant issue for the region, especially those countries closest to the Antarctic ozone hole such as Argentina and Chile. Following ratification of the Montreal Protocol, governments in collaboration with private sector and other stakeholders adopted regulations, established institutions and took measures to phase out ODS - Brazil halted production in 1999.

Countries such as Argentina, Mexico and Venezuela that still produce CFCs have developed policies and measures to reduce both production and consumption of ODS. In contrast to other developing regions, the Latin American and Caribbean Region has reduced total production of CFCs from the 1986 level by approximately 21 per cent. Global climate change may seriously affect the region given its ecological and socio-economic vulnerability. Changes in the water cycle may pose a danger to arid and semi-arid zones and thus affect the production of cereals and livestock as well as hydroelectric power generation in countries such as Argentina, Chile, Costa Rica and Panama.

Coastlines and coastal ecosystems in Central America, Argentina, Uruguay and Venezuela may be affected, and the coastal infrastructure may be damaged. Many of the large metropolitan areas are highly vulnerable to sea level rise, especially major ports. In the Caribbean, the small island states will probably be the first to suffer from sea level rise. There may also be impacts on health caused by an increase in epidemic disease vectors, along with other gastrointestinal infections.

Countries in the region do not have commitments under the UNFCCC or Kyoto Protocol. Mitigation and adaptation activities include energy-saving measures in transport, agriculture and waste management sectors, developing renewable energy sources and developing carbon sinks, mostly forests. Wind energy is exploited in several countries including Barbados, Costa Rica and Jamaica. A 2-MW demonstration plant using ocean thermal energy conversion has been built in Jamaica.

Over the past 30 years, there have been notable air quality improvements at both regional and local levels in the region. Levels of many air pollutants have been gradually reduced - the trends in the United States are representative of the region. Acid rain control programmes contributed to the dramatic decline in sulphur emissions since 1995 with reductions of 10-25 per cent in some parts of the north-eastern United States.

Ozone molecules have been found to travel large distances from emission sources - the typical transport range of tropospheric O_3 is 240-800 kilometres. Between 30 and 90 per cent of eastern Canada's O_3 comes from the United States, while the province of Ontario, the region in Canada that suffers from the worst O_3 problem, is a source of NO_x downwind into northeastern United States. Fossil fuel power plants are the largest point sources of NO_x - significant amounts of O_3 are formed and transported within the plumes of power plants.

In addition, while VOCs have decreased in the United States over the past 30 years, NOx emissions increased between 1970 and 1999 by 17 per cent. These findings led to a new approach in which North America recognized the need for aggressive strategies to reduce regional NOx emissions and for cooperation between the two countries. Under the Canada/US Air Quality Agreement, both countries set targets to reduce NOx emissions and in October 2000 they signed an annex to the agreement to reduce border emissions of NOx from fossil fuel power. They have also engaged in the 1995 North American Research Strategy for Tropospheric Ozone and signed the 1999 CLRTAP Protocol to Abate Acidification, Eutrophication and Ground-Level Ozone.

Recognition that exposure to ozone at concentrations below 0.08 ppm results in severe health effects has prompted revisions in both Canadian and US ozone health standards. Although levels of SPM have decreased by 40 per cent since 1980, recent research has revealed serious health concerns at concentrations well within allowed levels attributable to the finer airborne particles released mainly from vehicles and power plants. Consequently, North American standards for particulate matter have been adjusted.

North America's northern regions have been subject to serious stratospheric ozone depletion. With the 1987 Montreal Protocol on Substances that Deplete the Ozone Layer, both countries committed themselves to actions to protect the stratospheric ozone layer. In Canada, strict regulations passed in 1990 and reformulated in 1999 reduced production faster than the protocol required, from a high of 27 800 tonnes/year in 1987 to 900 tonnes/year in 1996. In the United States, the use and trade of ODS is controlled through a marketable permit system and a tax on ODS. The subsequent price increase of ODS encouraged the use of alternatives. Both countries reduced their non-essential CFC consumption to zero by 1996.

Since 1972 North America's climate has warmed considerably, reflecting a global trend. About half of the average rise in North America's surface temperature during the past century - more than 0.6 °C - occurred since the late 1970s. North America emits more greenhouse gas than any other region, accounting for around 5 per cent of the world's population but nearly 26 per cent of global anthropogenic emissions of CO2 in 1998. North America has one of the world's most energy-consuming economies.

The transportation sector is the largest source of CO_2 emissions, accounting for 30.1 per cent of Canada's emissions in 1995 while in 1993 cars and light trucks were responsible for more than 20 per cent of US CO2 emissions. In 1997, the US transport sector accounted for around 5 per cent of global anthropogenic CO_2 emitted and more than one-third of total world transportation energy use. Two sharp price shocks in the oil market in the 1970s helped to increase awareness that oil is not a renewable resource. Energy-saving standards for vehicle bodies, engines and fuel efficiency in new passenger cars were introduced in the 1970s and strengthened in the 1980s.

However, a combination of factors conspired to drive energy use up during the 1980s. Progress in total and per capita energy efficiency slowed and CO_2 emissions continued to rise. Renewed efforts subsequent to the UNFCCC commitments also failed to curb CO2 emissions in the 1990s. In 1998, emissions were 14 and 11 per cent above 1990 levels in Canada and the United States respectively. Renewable energy production from hydropower, wind, solar, biomass and geothermal sources is increasing but still contributes only a small fraction of energy needs, supplying about 7 per cent of US domestic energy demand in 2000.

In the transport sector, progress made in car fuel efficiency and emission controls has been partially offset by increases in the number of automobiles, in distances travelled, and a trend since 1984 toward lightduty trucks and sport-utility vehicles. Under the 1997 Kyoto Protocol, Canada agreed to reduce greenhouse gas emissions to 6 per cent and the United States to 7 per cent below 1990 levels between 2008 and 2012. However, in early 2001, the United States announced that implementing the Kyoto treaty would be too harmful to the economy and that it would pursue other ways of addressing climate change.

The level of industrialization in West Asia is low in comparison with Europe and the United States but population growth, urbanization and an increase in oilrelated industries and other industrial activities have resulted in air pollution 'hot spots'. In the major cities and industrial compounds of West Asia, concentrations of the main air pollutants often exceed WHO guidelines by a factor of two to five. The burning of fossil fuels is the main cause of atmospheric air pollution and the main source of anthropogenic CO_2 emissions.

It accounts for all West Asia's commercial primary energy production which increased from 665.5 million tonnes of oil equivalent in 1972 to 974.2 mtoe in 1997, while energy consumption increased from 27.0 to 229.5 mtoe over the same period. The main sources of air pollution in the Gulf Cooperation Council (GCC) countries are oil refineries, oil gathering centres, oil platforms, petrochemical and fertilizer plants, and motor vehicles.

In the Mashriq countries, outdated technologies especially in power generation plants, fertilizer plants, smelters and cement factories have caused deterioration of air quality not only in industrial sites but also in nearby settlements. Amongst the air pollutants emitted, SPM is of great concern with levels well above maximum allowable concentrations. The economic loss due to the impact of poor air quality on human health in Syria is estimated at about US$188 million per year. However, recent trends in West Asia, and especially in the GCC countries, are towards adopting cleaner production approaches in industry, especially in the large oil, petrochemical, fertilizer and metal industries.

The increasing number of vehicles, poor traffic management, ageing cars and congested roads in major cities add to the level of air pollution. Many vehicles are in poor condition and about 30 per cent are older than 15 years and produce significantly higher emissions of hydrocarbons and NOx than new ones. Furthermore, leaded petrol is still in use in many countries, compounding health problems in cities and along major highways.

Along with atmospheric pollution caused by human activities, seasonal sand and dust storms contribute to air pollution in West Asia in general and along the northern coasts of the Arabian (Persian) Gulf in particular. The dust storms absorb pollutants such as pesticides and can transport them for long distances with adverse effects on the environment, the economy and quality of life. It is estimated that the annual amount of dust fall-out along the coastal area of Kuwait may reach 1 000 tonnes/km2 with an overall mean concentration of 200 µg/m3.

Transboundary air pollution is an emerging issue in the region. Stricter measures and regulations to control emissions, to promote the use of modern and efficient technologies, and towards restructuring the price of energy resources must be taken to curb air pollution. Energy efficiency programmes in the power, petroleum, transportation, industrial, agriculture and residential sectors are needed to reduce energy consumption and associated emissions of greenhouse gases.

The West Asia states (with the exception of Iraq) acceded to the Vienna Convention and the Montreal Protocol and its amendments. All countries in the region are users but not producers of ODS and programmes regulating the import and use of ODS have been developed at both national and regional levels. Regulations for specifications and ODS emissions have been enacted, and ozone offices and coordinating committees have been established to monitor the activities of companies related to the consumption, handling and storage of these chemicals.

Companies are gradually phasing out the use of ODS and most countries have frozen the consumption of ODS as required by the Montreal Protocol. Further efforts are needed to phase out methyl bromide which is still consumed in Jordan, Lebanon and Syria. The West Asia region is likely to be affected by climate change. The Arabian Peninsula and a number of islands (such as Bahrain) will probably be prone to sea level rise.

Temperature variations and changes in rainfall patterns will affect water resources and food production capacity. Climate change impacts are considered a high priority in some countries, especially in island countries such as Bahrain. Following ratification of the UNFCCC, National Climate Change Committees were established and some countries have started to monitor air quality and meteorological parameters. National inventories of greenhouse gases have been completed for several countries and work is underway in other countries.

The key atmospheric issues in the Arctic and Antarctic are the depletion of the stratospheric ozone layer, the long-range transport of air pollutants and warming associated with global climate change. These problems are mainly due to anthropogenic activities in other parts of the world. Seasonal stratospheric ozone depletion over Antarctica, and more recently over the Arctic, has been one of the major regional environmental concerns since it was noticed in 1985.

The depth, area and duration of the Antarctic ozone hole has steadily increased, reaching an all-time high of around 29 million km2 in September 2000. In the Arctic, average yearly stratospheric ozone levels in the 1990s had declined by 10 per cent from the late 1970s, increasing the risk of snow blindness and sunburn. The recovery of the stratospheric ozone layer in the polar regions depends primarily on the implementation of the Montreal Protocol on the Substances that Deplete the Ozone Layer.

Therefore the efforts of nations to phase out the use of ODS, even though they are located far from the poles, are of the utmost importance. Natural ecosystems in polar regions have low adaptive capacity and are highly vulnerable to climate change. Climate change is almost certainly responsible for the decrease in extent and thickness of Arctic sea ice, permafrost thawing, coastal erosion, changes in ice sheets and ice shelves, and the altered distribution and abundance of species in polar regions.

Other impacts of the warming trend include a recorded 15 per cent increase in Arctic precipitation, increased storm episodes, earlier springs and a later onset of freezing conditions, and decreased marine salinity. Permafrost thawing can itself add to climate change problems - for example, emissions of methane from tundra may increase while reductions in the extent of highly reflective snow and ice cover will magnify warming. These effects may continue for centuries, long after greenhouse gas concentrations are stabilized, and may cause irreversible impacts on ice sheets, global ocean circulation and sea-level rise.

Northern Hemisphere, the Arctic is more exposed to anthropogenic air pollution than the Antarctic. Prevailing winds carry polluting substances -including heavy metals, POPs and sometimes radionuclides - into the Arctic where they can stay airborne for weeks or months and be transported over long distances. Over much of the Arctic, levels of certain types of pollutants are so high that they cannot be attributed to sources within the region; they come from much further south. Major sources of anthropogenic radionuclides in the Arctic include fall-out from nuclear tests, releases from nuclear fuel reprocessing plants, and fall-out from the 1986 Chernobyl nuclear power plant accident.

A significant increase of radioactivity in Arctic indigenous people was registered after the Chernobyl accident, particularly amongst those who consumed significant quantities of foods that concentrate radiocaesium, such as reindeer meat, freshwater fish, mushrooms and berries. The phenomenon was mainly observed in 1986-89 in Norwegian and Swedish Saami and up to 1991 in the indigenous population of the Kola Peninsula, in the Russian Federation. Since then the levels have been gradually falling back towards the pre-accident levels.

Within the Arctic, the Russian Federation's industrial complexes have been a major source of atmospheric pollution. Emissions of sulphur compounds and heavy metals from smelters have caused major forest degradation on the Kola Peninsula and have decreased the number of species in the region. The areas severely affected by air pollution around the Nickel-Pechenga and Varanger smelters increased from around 400 km2 in 1973 to 5 000 km^2 in 1988. Since 1990, emissions from Russian smelters have decreased or stabilized mainly because of the economic slowdown. The level of air pollution in the

Arctic is so high that 'Arctic haze' has become a major problem.

The term was coined in the 1950s to describe an unusual reduction in visibility that the crews of North American weather reconnaissance planes observed during flights in the high latitudes in the Arctic. The haze is seasonal, with a peak in the spring, and originates from anthropogenic sources of emission outside the Arctic. The haze aerosols are mainly sulphurous originating from coal burning in the northern mid-latitudes, particularly in Europe and Asia. The particles are about the same size as the wavelength of visible light, which explains why the haze is so apparent to the naked eye. Improvement in the state of the polar environment depends primarily on policies and measures implemented by people inside and outside of the polar areas.

The Arctic countries have taken a number of steps to improve air quality. These include signing the Convention on Long-Range Transboundary Air Pollution (CLRTAP) and the relevant protocols to it, and supporting the development of the Stockholm Convention on Persistent Organic Pollutants. In addition, domestic regulatory measures taken in the United States and Canada have reduced emissions of some POPs, heavy metals and sulphur compounds. Actions to address stratospheric ozone depletion rely on the successful implementation of the Montreal Protocol by all nations. Given the predicted increase in the global mean temperature, climate change will impose significant pressures on the polar regions in the 21st century.

These impacts are likely to be exacerbated by the high vulnerability and low adaptive capacity of polar ecosystems and of some traditional indigenous communities. Despite increasing activity in both the domestic and international arenas, only preliminary steps have been taken to address the problem of global climate change. The main regional challenge is therefore enhancing the potential for adaptation to change that should help mitigate adverse impacts.

8

Freshwater

The total volume of water on Earth is about 1,400 million km3 of which only 2.5 per cent, or about 35 million km3, is freshwater. The principal sources of water for human use are lakes, rivers, soil moisture and relatively shallow groundwater basins. The usable portion of these sources is only about 200 000 km3 of water - less than 1 per cent of all freshwater and only 0.01 per cent of all water on Earth. Much of this available water is located far from human populations, further complicating issues of water use.

About one-third of the world's population lives in countries suffering from moderate-to-high water stress - where water consumption is more than 10 per cent of renewable freshwater resources. Some 80 countries, constituting 40 per cent of the world's population, were suffering from serious water shortages by the mid-1990s and it is estimated that in less than 25 years two-thirds of the world's people will be living in water-stressed countries.

By 2020, water use is expected to increase by 40 per cent, and 17 per cent more water will be required for food production to meet the needs of the growing population. The three major factors causing increasing water demand over the past century are population growth, industrial development and the expansion of irrigated agriculture. Agriculture accounted for most freshwater withdrawal in developing economies in the past two decades. Planners have always assumed that growing demand would be met by taming more of the hydrological cycle through building more infrastructure. The damming of rivers has traditionally been one of the main ways to ensure adequate water resources for irrigation, hydropower generation and domestic use.

About 60 per cent of the world's largest 227 rivers have been strongly or moderately fragmented by dams, diversions or canals, with effects on freshwater ecosystems. This infrastructure has provided important benefits in the form, for example, of increased food production and hydroelectricity. There have also been major costs. Over the past 50 years, dams have transformed the world's rivers, displacing some 40-80 million people in different parts of the world, and causing irreversible changes in many of the ecosystems closely associated with them. Emphasis on water supply, coupled with weak enforcement of regulations, has limited the effectiveness of water resource management, particularly in developing regions.

Agriculture accounts for more than 70 per cent of freshwater drawn from lakes, rivers and underground sources. Most is used for irrigation which provides about 40 per cent of world food production. Over the past 30 years, the area of land under irrigation has increased from less than 200 million ha to more than 270 million ha. During the same period, global water withdrawals rose from about 2 500 km3 to more than 3 500 km3.

Poor management has resulted in the salinization of about 20 per cent of the world's irrigated land, with an additional 1.5 million ha affected annually, significantly reducing crop production. The countries most severely affected are mainly in arid and semi-arid regions. For many of the world's poorer populations, one of the greatest environmental threats to health remains the continued use of untreated water. While the percentage of people served with improved water supplies increased from 79 per cent (4.1 billion) in 1990 to 82 per cent (4.9 billion) in 2000, 1.1 billion people still lack access to safe drinking water and 2.4 billion lack access to improved sanitation. Most of these people are in Africa and Asia.

Lack of access to safe water supply and sanitation results in hundreds of millions of cases of water-related diseases, and more than 5 million deaths, every year. There are also large, but poorly quantified adverse impacts on economic productivity in many developing countries. The importance of meeting basic human needs for water has always played a major role in water policy. One of the earliest comprehensive water conferences was held in 1977 in Mar del Plata, Argentina. The focus on human needs led to the International Drinking Water Supply and Sanitation Decade and the efforts of the United Nations and other international organizations to provide basic water services.

The concept of meeting basic water needs was reaffirmed during the 1992 Earth Summit in Rio de Janeiro and expanded to include ecological water needs. A recent United Nations report recognized that all people require access to adequate amounts of safe water, for drinking, sanitation and hygiene. Most recently, the Second World Water Forum and Ministerial Conference in The Hague in 2000 produced a strong statement from more than 100 ministers in support of re-emphasizing basic human needs as a priority for nations, international organizations and donors.

Providing urban dwellers with safe water and sanitation services has remained a particular challenge. Some 170 million developing country urban dwellers were provided with safe water and 70 million with appropriate sanitation during the first half of the 1990s but this had limited impact because about 300 million more urban residents still lacked access to safe water supply, while nearly 600 million lacked adequate sanitation by the end of 1994. However, a major area of success in many developing countries is related to investments in wastewater treatment over the past 30 years which have 'halted the decline in - or actually improved - the quality of surface water'.

Water quality problems can often be as severe as those of water availability but less attention has been paid to them, particularly in developing regions. Sources of pollution include untreated sewage, chemical discharges, petroleum leaks and spills, dumping in old mines and pits, and agricultural chemicals that are washed off or seep downward from farm fields. More than half of the world's major rivers are 'seriously depleted and polluted, degrading and poisoning the surrounding ecosystems, threatening the health and livelihood of people who depend on them'. In the 1990s, many new efforts were made to monitor water quality and institute better policies and programmes.

Groundwater Resources

About 2 billion people, approximately one-third of the world's population, depend on groundwater supplies, withdrawing about 20 per cent of global water annually - much of it from shallow aquifers. Many rural dwellers depend entirely on groundwater. The issues of groundwater use and quality have until recently received far less attention than surface water, and data on groundwater stocks and flows are even less reliable. However, in Europe, much attention has been paid to groundwater quality because many settlements depend on such resources for water supply.

Generally, groundwater resources are vulnerable to a variety of threats, including overuse and contamination. When use exceeds natural recharge over a long period, groundwater levels drop. Parts of India, China, West Asia, the former Soviet Union, the western United States and the Arabian Peninsula are experiencing declining water tables, limiting the amount that can be used and raising the costs of pumping to farmers. Overpumping of groundwater can lead to salt-water intrusion in coastal areas.

Concern over growing problems related to groundwater resources has galvanized the international community, governments and other stakeholders to start addressing them. For example, the Second World Water Forum in March 2000 organized a special workshop on groundwater. Some of the recommendations arising from the workshop included the need to raise public awareness and 'improve information availability, quality and accessibility to stakeholders, technical specialists and policy-makers'.

Transboundary water management

Water is widely shared among nations, regions, ethnic groups and communities. Water has been used as a tool and weapon of conflict, access to water has been a source of dispute and contention, and major water development projects have led to violence and civil strife. But shared waters can also be a source of cooperation. This is particularly evident today with the increase in the number of initiatives related to river basin management regimes and institutions committed to bilateral and/or multilateral management of transboundary water resources.

This can be traced back to the 1966 Helsinki Rules which laid the foundation for international principles for shared watercourses and influenced many specific river treaties. The Rules were subsequently followed up by various international efforts, including particularly the work of the UN International Law Commission, which led in 1997 to the United Nations Convention on the Law of the Non-Navigational Uses of International Watercourses.

The impact of this new convention is already being felt with the 14-member Southern African Development Community (SADC) adapting many of its principles in its revised protocol on shared watercourses. The recognition of river basin organization over the past 30 years has also resulted in the establishment of the International Network of Basin Organizations (INBO) in 1996, while other initiatives include the 1998 International Conference on Water and Sustainable Development, which declared that 'a common vision of riverine countries is needed for the efficient management and effective protection of transborder water resources'. The conference's priority action programme emphasized the need to:

— facilitate the exchange of accurate and harmonized information among riverine countries;

— promote consultation at all levels, especially within pertinent international institutions and mechanisms; and

— define medium-range priority action programmes of common interest to improve water management and decrease pollution.

Water and Ecosystems

Water development projects during the 20th century have had significant impacts on freshwater ecosystems by eliminating marshes and wetlands, removing water for other uses, altering flows, and contaminating water with industrial and human wastes. In many rivers and lakes, ecosystem functions have been lost or impaired. In some areas, growing water demand has led to reductions in the volume of large rivers, affecting riverine and adjacent coastal areas.

Reproductive failures and death in various wildlife species, particularly at higher levels in the food chain, have been reported as a result of high withdrawals of water. Wetlands are an important freshwater ecosystem influencing not only species distribution and biodiversity in general but also human settlements and activities. They provide natural flood control, carbon storage, natural water purification, and goods such as fish, shellfish, timber and fibre.

While information on the real extent of wetlands globally remains unreliable, recent estimates show that they may cover at least 12.8 million km2. Human activities, including agriculture and settlements, have caused serious damage to freshwater ecosystems and contributed to the loss of about 50 per cent of the world's wetlands during the 20th century. This damage to ecosystems reduces water quality and quantity, leading to a reduction in the effective availability of water for human use. The total area of wetlands lost during the past 30 years is difficult to assess because of paucity of data and lack of accurate global information on the original extent of wetlands.

There has recently been a marked shift in water policy as policy-makers recognize that ecosystems require adequate water supplies to sustain normal functions and biodiversity conservation. Since 1992, new water policies have been developed that address the goal of preserving and allocating water for the environment - a change since the 1972 Stockholm Conference which focused mainly on protecting air and water quality, and paid no attention to water for ecosystems.

Although large projects involving dams are becoming less common due to limited sites, rising costs and general opposition, 349 dams more than 60 metres high were under construction in 1998. The remaining large free-flowing rivers are now found only in the tundra regions of North America and the Russian Federation, and in smaller basins in Africa and Latin America. There is now a growing emphasis on increasing wateruse efficiency and increased productivity with the limited resources available. Throughout the world there are great opportunities to meet human needs with less water by:

— using existing technology (such as drip irrigation, low-flow toilets, and better industrial processes);

— changing irrigation technology;

— finding and stopping wasteful leaks;

— changing wasteful practices (such as irrigating during the day, using potable water for landscape irrigation);

— charging proper prices for water; and

— changing human activities (shifting to more waterefficient crops, changing industrial processes away from water-intensive production).

Policies and institutions

The Commission for Sustainable Development (CSD) has reported that many countries lack adequate legislation and policies for efficient and equitable allocation and use of water resources. Progress is, however, being made with the review of national legislation and enactment of new laws and regulations. Concern has also been expressed about the growing incapacity of national hydrological services and agencies, particularly in developing countries, to assess their respective water resources.

Many agencies have experienced reductions in observation networks and staff despite increases in water demand. A number of response measures have been undertaken including the World Hydrological Cycle Observing System (WHYCOS), which has been implemented in several regions. Its main objective is to contribute to the improvement of national and regional water resource assessment capabilities. Many different kinds of organizations play a role in water policy decisions, from national governments to local community groups.

Over the past decades, however, there has been a growing emphasis on increasing the participation and responsibility of small, local groups and an acknowledgement that communities have an important role to play in water policy. The Ministerial Declaration at The Hague in March 2000 called for 'Governing water wisely: to ensure good governance, so that the involvement of the public and the interests of all stakeholders are included in the management of water resources'. The private sector has recently begun to expand its role in water management.

The 1990s saw a rapid increase in the rate and extent of privatization of previously publicly managed water systems. Private water companies are increasingly serving the needs of growing cities by taking over contracts from public agencies to build, own and operate some or even all of a municipal system. At the same time, concerns have been growing about how best to ensure equitable access to water for the poor, finance projects and share risks.

Africa's renewable water resources average 4 050 km3/year, providing in the year 2000 an average of about 5 000 m3 per capita/year - significantly less than the world average of 7 000 m3 per capita/year and less than one-quarter of the South American average of 23 000 m3 per capita/year. However, the distribution of both surface water and groundwater is uneven. The spatial distribution of water resources in the region does not coincide with the highest population densities, resulting in many areas being water stressed or dependent on external sources of water.

Groundwater is a major source of water in the region, contributing 15 per cent of Africa's resources. The major aquifers are found in the northern Sahara, Nubia, Sahel and Chad basins as well as the Kgalagadi. Groundwater is used for domestic and agricultural consumption in many areas, particularly the more arid sub-regions where surface water resources are limited. However, areas heavily dependent on groundwater reserves are also at risk of water shortages, as water is extracted far more rapidly than it is recharged.

Africa experiences large spatial variations in rainfall, with 95 per cent of the total falling in the central and southwestern wet equatorial zone. Severe droughts have been recorded in the Sahel and in the Northern, Eastern and Southern subregions over the past 30 years. As a result, a number of inter-basin water transfer schemes have been developed. For example, in South Africa where 60 per cent of the run-off originates from one-fifth of the land area, large volumes of water are transported by inter-basin transfer schemes to major industrial centres such as Johannesburg.

However, these schemes can exert significant pressure on the environment as reduced natural flows impact on downstream ecosystems. Responses to water shortage in the Seychelles and Mauritius include desalinization, water rationing by the hotel and manufacturing industries, and recycling of domestic wastewater. These measures are expected to produce savings of about 240 million m3/year in the long-term.

In Egypt, severe water shortages have also prompted schemes to recycle agricultural drainage water to meet growing agricultural demand. As in other regions, the major factors influencing water availability in Africa include growing domestic consumption for drinking water and sanitation, irrigated agriculture and industrialization. Losses from domestic water distribution systems account for significant wastage. Similarly, many irrigation systems are outdated. In South Africa, up to 50 per cent of irrigation water is lost due to leakages.

Some 62 per cent of Africans had access to an improved water supply in 2000. Even so, rural Africans spend much time searching for water and 28 per cent of the global population without access to improved water supplies live in Africa. Women are particularly affected as they are often responsible for the family's water needs. Urban areas are better supplied, with 85 per cent of the population having access to improved water supplies. In rural areas, the average is 47 per cent, with 99 per cent of the rural population in Eritrea having no sanitation coverage.

The total African population with access to improved sanitation was 60 per cent in 2000. Again, urban populations fared better, with an average 84 per cent having improved sanitation compared to an average 45 per cent in rural areas. Poor water supply and sanitation lead to high rates of water-related diseases such as ascariasis, cholera, diarrhoea, dracunculiasis, dysentery, eye infections, hookworm, scabies, schistosomiasis and trachoma.

About 3 million people in Africa die annually as a result of water-related diseases. In 1998, 72 per cent of all reported cholera cases in the world were in Africa. Poor water supply and sanitation lead to contamination of surface and groundwater, with subsequent effects on plant, animal and human communities. The economic costs can be high. In Nigeria, a recent study estimates the future cost of water supply and environmental sanitation to be US$9.12 billion during 2001-10. Governments are trying to improve the situation with environmental management policies that include waste management and urban planning, and by making environmental impact assessments compulsory for large projects.

Freshwater and groundwater pollution is a growing concern in many areas, further limiting access to safe water. Poor water quality leads not only to waterrelated diseases but also reduces agricultural production, which means that more foodstuffs and agricultural products must be imported. Poor water quality also limits economic development options, such as water-intensive industries and tourism, a situation that is potentially disastrous to developing countries in Africa.

To counter this problem, many countries have established or enforced effluent water standards and rehabilitated wastewater treatment facilities. Other responses include schemes in Central Africa for purification and decontamination of freshwater systems, and public awareness campaigns. Although only recently implemented, these responses have been locally successful in improving access to drinking water and raising awareness.

In Eastern and Southern Africa, the widespread invasion of the water hyacinth is a further cause of deteriorating water quality. The water hyacinth forms dense mats that block water channels, disrupting flow patterns. Decaying mats of the weed generate bad odours and lead to eutrophication of the water body. Areas afflicted by the water hyacinth include Lakes Victoria and Kariba, and some rivers. Affected countries have begun biological and chemical control programmes, in addition to mechanical clearance, with some success.

Wetland habitats in Africa cover about 1.2 million km2. However, wetlands are threatened by both pollution and reclamation. Loss of wetlands in Southern Africa is thought to have contributed to the severity of the flooding in 1999-2000, which affected 30 000 families and 34 000 hectares of farmland. To prevent further wetland degradation, 27 African countries had signed and ratified the 1987 Ramsar Convention as of December 1998, placing 75 sites covering some 14 million ha under protection.

Integrated water resources management A shift towards integrated water resources management (IWRM) is one of the new policy initiatives that have been adopted in Africa to address the issues raised above. Integrated water resources management is not limited to the national level but also includes basins shared by two or more states. The Nile Basin Initiative, launched in 1999, is a joint programme of action between 10 Nile countries. The objectives are to ensure sustainable resource development, security, cooperation and economic integration.

In Southern Africa, the eight basin states of the Zambezi have been cooperating under the Zambezi River System Action Plan but efforts to establish a Zambezi Basin Commission have been slow. Another example of regional cooperation is in the Lake Victoria area, where in 1995 a GEF-funded project was established to focus primarily on fisheries management, pollution control, control of invasive weeds, and catchment land use management.

The Asia and the Pacific Region accounts for about 36 per cent of global run-off. Even so, water scarcity and pollution are key issues and the region has the lowest per capita availability of freshwater: renewable water resources amounted to about 3,690 m3 per capita/year in mid-1999 for the 30 largest countries in the region for which records are available. In absolute terms, China, India and Indonesia have the largest water resources, more than one-half of the region's total. Several countries, including Bangladesh, India, Pakistan and the Republic of Korea, already suffer from water scarcity or water stress.

Water Scarcity

Many countries do not have sufficient water to meet demand, with the result that aquifer depletion due to overextraction is common. Moreover, the scarcity of water is accompanied by a deterioration in the quality of available water due to pollution and environmental degradation. Dams and reservoirs coupled with deforestation in some watersheds have reduced stream water levels, lowered water tables, degraded riparian wetlands and diminished freshwater aquatic diversity. Excessive demand for groundwater in coastal cities such as Bangkok, Dhaka, Jakarta, Karachi and Manila has led to saline intrusion and ground subsidence.

Government policies and strategies have traditionally been concerned with increasing supply. However, policies have recently become increasingly focused on an integrated approach to water resource management by emphasizing demand management measures such as efficient water use, conservation and protection, institutional arrangements, legal, regulatory and economic instruments, public information and inter-agency cooperation. Common elements in the national policies and strategies now being adopted include integration of water resources development and management into socio-economic development; assessment and monitoring of water resources; protection of water and associated resources; provision of safe drinking water supply and sanitation; conservation and sustainable use of water for food production and other economic activities; institutional and legislative development; and public participation.

In India, a new irrigation management policy aims to improve water application efficiency through the use of modern technologies such as drip/sprinkler irrigation and better on-farm irrigation measures. In the Republic of Korea, where agriculture uses more than 50 per cent of water resources, the government's water resources development plan for the 21st century highlights measures that relate to increased food production with efficient water use. Decentralized water management is also being encouraged in countries such as China where city or provincial authorities are authorized to manage water resources.

In India, multidisciplinary units in charge of developing comprehensive water plans have been established in some states. Stakeholder participation has reduced operational costs in countries such as Pakistan by involving communities in the development of water supply, sanitation, and water pollution prevention facilities and their maintenance. Progress has also been made in adopting a basinwide approach. The Indus Basin water-sharing accord between India and Pakistan, the acclaimed Water Sharing Treaty between India and Bangladesh, the India-Bhutan cooperation on hydropower development and India-Nepal cooperation in harnessing transboundary rivers are examples of transboundary cooperation on water management in South Asia. A major challenge is to change the fragmented sub-sectoral approaches to water management that have caused conflict and competition in the past, and to design and implement integrated mechanisms, particularly for projects that transcend sub-sectors.

Water pollution

Over the years, water pollution has emerged as a major issue. Pollutants include pathogens, organic matter, nutrients, heavy metals and toxic chemicals, sediments and suspended solids, silt and salts. South Asia - particularly India - and Southeast Asia are facing severe water pollution problems. Rivers such as the Yellow (China), Ganges (India), and Amu and Syr Darya (Central Asia) top the list of the world's most polluted rivers. In cities in the developing countries of the region, most water bodies are now heavily polluted with domestic sewage, industrial effluents, chemicals and solid wastes.

Most rivers in Nepal's urban areas have been polluted and their waters are now unfit for human use, while drinking water in Kathmandu is contaminated with coliform bacteria, iron, ammonia and other contaminants. Water pollution has affected human health. In the Pacific Islands, especially in some atoll communities, use of polluted groundwater for drinking and cooking has led to health problems such as diarrhoea, hepatitis, and occasional outbreaks of typhoid and cholera.

Groundwater in districts of West Bengal, India, and in some villages in Bangladesh, for example, is contaminated with arsenic at levels as much as 70 times higher than the national drinking water standard of 0.05 mg/litre. While pollution is a factor, arsenic contamination is also due to natural phenomena. According to one report, 'With the majority of the country's 68 000 villages potentially at risk, UN scientists estimate that the arsenic may soon be killing 20,000 Bangladeshis a year'.

Inadequate water supply and poor sanitation cause more than 500 000 infant deaths a year as well as a huge burden of illness and disability in the region. Some 8-9 per cent of the total Disability Adjusted Life Years (DALYs) are due to diseases related to inadequate water supply and poor sanitation in India and other countries.

Cholera is prevalent in many countries, particularly those where sanitation facilities are poor such as Afghanistan, China and India.

Of the global population without access to improved sanitation or water supply, most live in Asia. In the Southwest Pacific sub-region, water supply and sanitation appear to be relatively good, with 93 per cent of the population having access to improved sanitation and 88 per cent to improved water supply. These figures are strongly biased by the large and well-served population of Australia, however.

During the past decade, several countries have started to address the water quality problem by implementing large-scale programmes and action plans to rehabilitate degraded streams and depleted aquifers. These programmes are typically given legislative or statutory authority such as that provided by Thailand's National Water Quality Act, the Philippine Water Quality Code, India's Environment Protection Act, China's Water Law and the Republic of Korea's Water Quality Preservation Act.

Clean-up campaigns for rivers, canals, lakes and other water bodies have become widespread. The programmes have often been successful in improving water quality and, occasionally, have led to the adoption of new water quality standards and water use regulations. They have also increased awareness of the need to reduce pollutant loads through wastewater treatment, reuse and recycling of sewage and industrial wastewater, introduction of low-cost technologies, and strict control of industrial and municipal effluent.

There have been a number of successes in water reuse and recycling in the industrialized countries of the region.

Water quality has been improved in China, Japan, the Republic of Korea and Singapore as a result of initiatives to address water pollution. In Japan, the government has set environmental quality standards and made remarkable improvements: in 1991, 99.8 per cent of water samples met standards for heavy metals and toxins in Japan. In 2000, the rate of industrial wastewater treatment across China was 94.7 per cent. Action in Singapore means that Singaporeans can now enjoy drinkable piped water straight from the tap.

Water resources are unevenly distributed in Europe. Annual average run-off ranges from 3 000 mm in western Norway to 100-400 mm over much of central Europe and less than 25 mm in central and southern Spain. Traditionally, most European countries rely more on surface water than on groundwater, which is often used only for public water supply. Water pollution is a serious issue throughout Europe. Though some progress has been made in pollution reduction in Western Europe, the situation is less promising in Central and Eastern Europe (CEE).

Europe abstracts a relatively small portion of its total renewable water resources. Western Europe abstracts, on average, around 20 per cent, although this ranges from less than 5 per cent in the northern countries to more than 40 per cent in Belgium, Germany and the Netherlands. The Russian Federation, with 9 per cent of the world's water resources, abstracts less than 2 per cent annually. However, water quantity problems do occur in areas with low rainfall and high population density, and in large land areas under irrigated agriculture, especially in Mediterranean countries and in CEE.

In the central part of Western Europe, most of the water supply is used as a coolant in energy production. This water is returned to its source more or less unchanged and can be re-used. In southern countries of Western Europe, where water resources are less abundant, agriculture consumes far more than the other sectors - about 80 per cent, compared to 20 per cent for urban and industrial use. Some 80 per cent of the water used for irrigation evaporates. The amount of water abstracted for public water supply in Western Europe fell by 8-10 per cent between 1985 and 1995 as a result of improved industrial and domestic water use efficiency.

However, a significant increase in agricultural water use occurred in southern Europe because irrigated farmland has increased by nearly 20 per cent since the mid-1980s. In CEE, there has been a marked decline in water consumption for industrial purposes due to economic restructuring but the demand from urban areas and for irrigated agriculture is growing steadily. There is little legislation controlling water use in Europe. Traditionally, quantity problems have been dealt with by increasing storage capacities through reservoirs and water transfer schemes.

However, demand reduction measures are now in place in several countries in Western Europe. These, together with greater awareness of water use in general, have reduced public consumption. The domestic and industrial sectors have become increasingly waterefficient. Examples of water conservation measures include: metering; increased charges and taxes; restrictions on garden watering; reducing leakage; user education and widespread use of more efficient appliances such as low and dual flush toilets, and washing machines that consume less water.

Abstraction charges and pricing mechanisms are valuable for improving the sustainability of water use in agriculture and deserve more attention because water prices for agricultural use are often lower than for other users. For instance, a recent study found that in the French Adour-Garonne basin, the drinking water supply is almost entirely selffinancing but the irrigation tariff covers only 30-40 per cent of the total cost of the services. Other agricultural reforms include planting crops with lower water demands and introducing more efficient irrigation systems.

Overloading with organic matter, nitrogen and phosphorus in the 1970s and 1980s resulted in eutrophication of seas, lakes, rivers and groundwater throughout Europe. The main source of nitrogen is fertilizers in run-off from agricultural land. Most phosphorus comes from households and industry wastewater, though in areas with intensive agriculture in Western Europe, phosphorus from agriculture approaches 50 per cent of the total load.

In Western Europe, fertilizer consumption has fallen since the mid-1980s but eutrophication has continued due to increased nutrient run-off from intensive livestock production. In CEE, use of agrochemicals has declined markedly since the early 1990s, resulting in a reduction of nitrogen-phosphorus fertilizer use by about 50 per cent. Pollution of groundwater is another serious problem, mainly associated with nitrates and pesticides from agriculture.

In the Russian Federation alone, more than 2,700 sources of groundwater were identified as polluted in 1999. Phosphorus discharges from urban wastewater treatment plants in Western Europe have fallen significantly since the early 1980s, largely due to the huge increase in treatment of wastewater and wide-scale introduction of phosphorus-free detergents. By the end of the 1990s, 90 per cent of Western Europeans were connected to sewers and 70 per cent to wastewater treatment plants.

Since 1990, most Accession Countries have started to invest heavily in sewage collection and treatment but its high cost is one of the major financial issues in the accession process. In the Eastern European countries of the former Soviet Union, little has been done to improve wastewater treatment. Many lakes that had high phosphorus concentrations in the early 1980s have lower concentrations today. However, only slight changes in phosphorus concentrations have been observed in initially less-affected lakes.

This is mainly due to accumulation and (delayed) release of phosphorus from lake bottoms or continued contamination from small, scattered dwellings and from agricultural sources. Overall, water quality in many European lakes is still poor. Heavy pollution in Western European rivers such as the Rhine has declined significantly since 1980 but improvements have been less significant in southern and Central Europe.

In Eastern Europe, the situation is different. In the Russian Federation and Ukraine, the two most industrialized countries of the former Soviet Union, discharge of polluted water into rivers increased in the second half of the 1980s and in the 1990s, despite an alleged clean-up campaign for the Volga and Ural rivers as early as 1972. Poor water quality impacts human health. In Europe, however, outbreaks of water-borne diseases affecting less than 20 per cent of the supplied population are rarely detected. Even so, occasional outbreaks of water-borne diseases such as gastrointestinal infections, affecting much of the population, are reported across Europe, even from countries with high standards of supply.

There are many multilateral and bilateral agreements for the management of transboundary waters. At the pan-European level, the 1992 UNECE Convention of the Protection and Use of Transboundary Watercourses and International Lakes strengthens national measures, obliging parties to prevent, control and reduce water pollution from point and non-point sources. It also includes provisions for monitoring, research and development, consultations, warning and alarm systems, mutual assistance, institutional arrangements, and the exchange and protection of information, as well as public access to information.

A Protocol on Water and Health is awaiting entry into force. At the watershed level, transboundary initiatives include the Convention on Cooperation for the Protection and Sustainable Use of the River Danube and the new Convention for the Protection of the Rhine. The Danube Convention pledges the signatories to work together to conserve, improve and rationally use the surface and ground waters in the Danube Catchment basin; to control hazards originating from accidents in the river area; and to contribute to reducing the pollution loads of the Black Sea from sources in the catchment area.

The new Rhine Convention, adopted at the January 2001 Conference of Rhine Ministers, will be the basis for international cooperation of the riparian countries and the EU, replacing the Agreement on the International Commission for the Protection of the Rhine against Pollution and the 1976 Convention for the Protection of the Rhine against Chemical Pollution. The new convention fixes targets for international cooperation for sustainable development of the Rhine, further improvement of its ecological state, holistic flood protection and defence.

The Latin America and Caribbean Region is rich in renewable water resources with more than 30 per cent of the world's total. However, three hydrographic regions - the Gulf of Mexico Basin, the South Atlantic Basin and the La Plata Basin - covering 25 per cent of the region's territory, are home to 40 per cent of the population and contain only 10 per cent of the region's water resources. Most water-related problems transcend national boundaries although there are marked differences between sub-regions and countries. The major challenges are: decreasing per capita water availability due to population growth, urban expansion, deforestation and climate change; deteriorating water quality arising from untreated sewage, excessive use of fertilizers and pesticides, and industrial pollution - particularly from the mining and energy industries; and outdated institutional and legal frameworks.

Water availability varies greatly, with South America being the richest sub-region. Some countries suffer from water shortages, and loss of aquatic biodiversity and habitats, with some Caribbean islands approaching or below water scarcity levels. Sharp differences also exist within countries. In most Caribbean small island states, rainfall is the sole source of freshwater. In South America, groundwater reserves are of great importance and are estimated at 3 million km3.

Mexico is particularly dependent on groundwater, which accounts for one-third of all reported freshwater extraction and for two-thirds of urban drinking water. Agriculture and industry are the major consumers of water in the region, followed by domestic consumption. Irrigated agriculture is one of the fastest expanding uses of freshwater. The amount of land under irrigated agriculture increased from 10 million ha in 1970 to more than 18 million ha in 1998. Water withdrawals for irrigation range from 56 per cent of total withdrawals in the Caribbean to 78 per cent in Mesoamerica. There is a general lack of efficiency in irrigation technology and practice. Some recent institutional reforms have attempted to address this.

In Mexico, for example, ownership of public irrigation systems has been transferred to 386 Water User Associations resulting in a dramatic improvement in cost recovery, system maintenance, yield and water use efficiency. Industry also uses large amounts of water. In South America, it is estimated that annual industrial withdrawals reach 15 km3 with 80 per cent of this demand coming from Argentina and Brazil alone. In Brazil, virtually all the country's electricity comes from hydropower. The mining sector, especially in Chile and Peru, requires increasing quantities of water.

For some regions such as the Andes, this may well lead to the need to import water in the near future. In Venezuela and Trinidad and Tobago, the petroleum sector is an important consumer. The demand for water for domestic use is also on the increase. However, inequity among users, even in water-rich countries, is rampant. Many of the poor in both rural areas and urban communities have neither access to clean water nor sanitation services. In 1995, 27 per cent of the population had either no domestic water supply or no easy access to it. In the same year, 41 per cent of the water was untreated and 31 per cent of the population had no sewage services.

By 2000, 85 per cent of the population had an improved water supply and 78 per cent had improved sanitation coverage - but this still meant that 78 million people had no access to improved water supply and 117 million had no access to improved sanitation services. Large disparities also exist between rural and urban areas. Natural disasters provide an additional, unexpected factor that can seriously undermine efforts to improve water and sanitation services. Attempts have been made to improve water supply and sanitation in many urban areas and to set tariffs that reflect the real value of water.

Although the efficiency of privatization and the use of economic instruments such as water pricing remain highly contentious, some countries, such as Jamaica, have started using economic instruments. Limited information on water utility infrastructure and operation is a major constraint in government efforts to improve regulations related to the urban water sector. While the role of governments has shifted from provider of a water service to regulator representing the public, many governments still do not have sufficient information on water utility operation, and this limits their regulatory functions.

Although the rate of technological change in the water sector is generally slower than in other sectors, the need to transfer technology is critical to efforts to save water and introduce improved control strategies. To increase the efficiency of the sanitation sector, as well as to attract capital, national and international initiatives have been suggested, including the creation of local or regional markets such as the Guarani Aquifer project. In Brazil, there have been notable advances in legislation, especially with a 1997 federal law that put in place a national water resources policy and established a national water resources management system.

Water pollution problems in Latin America and the Caribbean did not become a serious issue until the 1970s. Over the past 30 years, however, there has been a significant decrease in the quality of surface water and groundwater. Agriculture and the release of untreated urban and industrial sewage have become the major sources of pollution. The excessive use of fertilizers in agriculture has enhanced algal growth and eutrophication in lakes, dams and coastal lagoons. Rising levels of nitrates have been observed in rivers, including the Amazon and the Orinoco, as well as in underground sources in the region.

In Costa Rica, levels of nitrates approaching or surpassing international guidelines have been found in both metropolitan and rural sources. Untreated sewage from urban centres remains a major cause of pollution. In the region as a whole, only about 13 per cent of collected sewage receives any kind of treatment. Increasing pollution from urban run-off and the release of untreated sewage into water bodies serving urban areas has added to the difficulties of meeting the increasing demand for water in cities, especially those located in areas such as Lima and Mexico City where water is scarce.

Industrial activities, and the resulting pollution, have contributed substantially to water quality problems. Animal waste from tanneries, slaughterhouses and meat packing plants, for example, pollutes aquifers with coliform bacteria. Another water quality problem that is becoming more prevalent, especially in the Caribbean, is the salinization of water in coastal areas due to overextraction. This is particularly severe given the rising demand for water in the Caribbean, mainly to service the tourism industry.

In most countries, water resources continue to be managed on a sectoral basis with little integration either between sectors or with other environmental management procedures. Such an approach ignores vital interactions with much wider ecosystems and other functions, and ecological services related to water. There has been a trend towards transferring water services from the public to the private sector during the past decade and to decentralizing legal and administrative responsibilities. As a result, laws and regulations designed to protect freshwater resources are often lacking or poorly enforced.

North America holds about 13 per cent of the world's renewable freshwater (excluding glaciers and ice caps). At the end of the 1990s, North Americans used 1 693 cubic metres of water per person per year, more than in any other region. In the United States, recent conservation measures have led to declines in consumption: during 1980-95, water withdrawals declined by nearly 10 per cent while the population increased by 16 per cent. In Canada, on the other hand, water withdrawal increased by 80 per cent during 1972-91 while the population grew by 3 per cent.

Although point source water pollution has been reduced in the United States since the 1970s, nonpoint sources, such as agricultural run-off and urban storm drainage, have grown causing serious pollution problems. Nutrient enrichment problems are of particular concern. Most of the continent's freshwater resources lie in groundwater. Groundwater contamination and declining aquifer levels are now priority issues. By the mid-1990s, groundwater was supplying up to 50 per cent of the North American population and more than 90 per cent of rural dwellers.

The many hazardous compounds used in industry and agriculture are now threatening groundwater quality. Contaminants from non-point sources are present in many shallow wells throughout large regions of North America. Agriculture is the worst offender, with artificial fertilizer use in the region increasing from 15 to 22.25 million tonnes a year over the past 30 years. Although nitrogen contamination rarely exceeds levels of potential health risk, it is a chronic problem for the population in the Prairie Provinces that rely on wells for water, and it affects groundwater to some extent in 49 US states. Consumed in high concentrations, nitrates can cause infant methaemoglobinaemia or blue-baby syndrome.

During 1993-95, low concentrations of pesticides were also detected in shallow groundwater in 54.4 per cent of US sites tested. Although concentrations of pesticides rarely exceed drinking water standards, some scientists suggest that their combined effects on health and the environment are not adequately assessed. Underground storage tanks containing, for example, petroleum products, acids, chemicals and industrial solvents, are leading sources of groundwater contamination. The tanks are often inappropriate containers for these substances or have been improperly installed.

The long-term availability of groundwater in arid agricultural regions is a priority issue. In general, groundwater levels stopped declining during the 1980s but depletion of stored groundwater still accounted for about 10 per cent of all freshwater withdrawals in the United States in the mid-1990s. Agriculture relied on groundwater resources for 62 per cent of its irrigated farmland in 1990. During the late 1980s and early 1990s, all US states enacted groundwater legislation. The Canadian federal government has initiated new national legislation on the environment, trade and groundwater issues.

Although groundwater management has traditionally focused on surface and groundwater separately, interactions between them have direct effects on water quality and availability, and on the health of wetlands, riparian ecology and aquatic ecosystems in general. The Great Lakes basin is one of the Earth's largest freshwater systems, containing 18 per cent of the world's fresh surface water. Less than 1 per cent of the water is renewed annually by precipitation, surface water run-off and groundwater inflow.

Over the years, the lakes have been subject to a polluting mix of effluents due to inadequate sewage treatment, fertilizer and wastewater effluent. By the early 1970s, beaches were smothered with algae and water was unfit for drinking unless extensively purified. Lake Erie suffered from excess phosphorus, algal blooms and serious declines in fish populations. Aboriginal communities were the most affected.

In the early 1970s, eggshells of the double-crested cormorant, which is high on the aquatic foodchain and subject to the effects of bioaccumulation, were some 30 per cent thinner than normal. Some species of bird populations crashed. The International Joint Commission (IJC) released a report on the pollution problem in the lower Great Lakes in 1970. The IJC, an independent organization of Canadian and US representatives, has been in charge of assessing water quantity and quality along the boundary between Canada and the United States since 1909.

Rapid urban and industrial development, however, continued to cause environmental damage to the watershed during the 1990s. Sediment contamination in harbours and river mouths threatened to contaminate fish and posed problems related to dredging and sediment disposal. The Great Lakes Binational Toxics Strategy was launched in 1997 to eliminate these chemical contaminants. Although exposure to persistent toxic contaminants has decreased, some studies show that children of mothers who ate large quantities of Great Lakes fish had development problems.

The Great Lakes will face other environmental challenges in the future. Global warming could lower lake levels by a metre or more by the middle of this century, causing severe economic, environmental and social impacts. Water shortages throughout North America may also increase pressure to divert or remove water in bulk from the lakes, threatening the sustainable use of surface and groundwater resources.

The Arabian Peninsula is characterized by an arid climate with annual rainfall of less than 100 mm. There are no reliable surface water supplies. The subregion depends entirely on groundwater and desalination plants to meet its water requirements. Large increases in demand have placed the meagre resources available under increasing pressure. The Mashriq sub-region is mostly arid and semi-arid. About 70 per cent of the sub-region receives less than 250 mm of rain a year. The Mashriq has two shared rivers originating outside the area, the Euphrates and Tigris, and many smaller ones. Agreements or understandings about how to share these water resources have been reached among Arab countries but agreements over the Euphrates are yet to materialize between Iraq and Syria on one side and Turkey on the other.

The major cause of the increasing demand for water is rapid population growth. In many countries, water rationing is used to limit demand. For example, Jordan restricts water supplies in Amman to only three days a week. In Damascus, water can be used for less than 12 hours a day. Agriculture is the main user of water in West Asia, accounting for nearly 82 per cent of the total water consumed compared to 10 per cent and 8 per cent for the domestic and industrial sectors, respectively. In the Arabian Peninsula, agriculture utilizes about 86 per cent of the available water resources, and about 80 per cent in the Mashriq. To satisfy water demand, especially for irrigation, groundwater abstraction has increased dramatically during the past three decades.

The water stress index in West Asia is more than 100 per cent in five of the seven countries in the Arabian Peninsula, and is critical in the remaining two. These countries have already exhausted their renewable water resources and are now exploiting non-renewable reserves. In the Mashriq, except in Jordan, the water stress index is lower. While per capita water resources in 9 of the 12 countries in West Asia are below 1 000 m3/year, they are also below 500 m3/year in seven countries.

Over the past three decades, the adoption of food self-sufficiency policies has encouraged agricultural expansion. Governments offered subsidies and incentives which resulted in a large-scale expansion of farming, increasing water demand which was satisfied mainly by mining deep aquifers. Furthermore, unregulated pumping, absence or minimal irrigation water tariffs, lack of enforcement measures against unlawful drilling, poor irrigation practices and lack of farmer awareness have resulted in excessive water usage.

Intensive agriculture and heavy application of agrochemicals have also contributed to the contamination of water resources. For example, the concentration of nitrates in Gaza's tap water exceeds WHO guidelines and nitrate concentrations are increasing at a rate of 0.2-1.0 mg/litre per year in the country's coastal wells. Adherence to WHO standards would place half of these coastal wells off limits as drinking water. Excessive use of groundwater has resulted in sharp declines in groundwater levels and quality deterioration due to seawater intrusion.

In the United Arab Emirates, excessive groundwater pumping has created cones of depression 50-100 km in diameter in several areas. These cones have caused groundwater levels to fall, shallow wells to dry up, and saltwater intrusion. Water quality degradation is often a consequence of both water scarcity and overexploitation. Water quantity and quality are both major issues in the Mashriq countries. Effluent, agrochemicals and industrial discharges have seriously affected aquatic life, causing public health hazards.

West Asia is developing policies to increase both water supply and conservation. In Jordan, priority is given to the sustainability of water resources without mining groundwater resources; the country is constructing dams and facilities to store all available water resources. Many countries have started to invest in more efficient irrigation technology. Wastewater reuse is another important conservation tool for non-potable uses, including irrigation, and for reducing environmental degradation and improving public health.

In the past three decades, water authorities in the region have concentrated their efforts on supply augmentation and, to a lesser extent, on demand management and conservation. Although their effectiveness is not yet proven, programmes related to demand management, conservation and protection have been implemented in both sub-regions. These programmes include the reduction of fuel and agricultural subsidies, metering of groundwater wells, future plans for an irrigation water tariff, subsidies for modern irrigation techniques, and public awareness campaigns.

In the GCC countries, these policies have been only partially successful in alleviating water scarcity caused by increasing demand and food self-sufficiency policies. In fact, the food self-sufficiency policies of the past three decades have not been successful. The deficit in food production is growing and is aggravated by the scarcity of land and water resources that are already overexploited. Water security will become one of the major constraints to further development in the region in the next 30 years unless there are major changes to agricultural and water policies.

The Arctic holds much of the world's freshwater supply and its landscape is dominated by freshwater systems. The two main permanent ice fields are the ice pack of the Arctic Ocean and the Greenland ice cap, which together hold 10 per cent of the world's freshwater. The Greenland ice cap produces about 300 km3 of icebergs a year. The Arctic has several of the world's largest rivers. Low temperatures, low nutrient, short light availability and a brief growing season limit the primary productivity of Arctic freshwater systems. This in turn restricts the animal life that can be supported.

Nevertheless, the river systems are heavily populated by several fish species such as the Arctic charr, and the North Atlantic and Pink salmon. In recent years, the overall warming trend plus increased recreational and commercial fisheries use have put pressure on these populations. Accidental introduction of alien species and increased fish farming is another source of concern. Eutrophication is a recent problem in several lakes in Scandinavia where human settlements have raised nutrient levels.

Northbound rivers are major pathways of pollutants from sources far inland, especially in the Russian Federation. In the spring, these contaminants are deposited into the freshwater systems and eventually into the marine environment and can be transported thousands of kilometres from their sources via the Arctic's marine circulation patterns. Contaminants include chemicals from agricultural, industrial and petroleum production, radionuclides from nuclear testing and military activities, and water soluble salts.

The Arctic countries adopted a circumpolar Regional Programme of Action for Protection of the Arctic Marine Environment against Land-based Activities as well as National Programmes of Action in some countries, including the Russian Federation. These instruments are too recent to assess long-term effectiveness. Opposition to damming is strong in the Nordic countries. During 1975-2001, the Cree people fought the government of Quebec over environmental damage to their lands. In a surprise move in October 2001, however, the Cree reversed their stand and signed an agreement in principle to allow the government of Quebec to build another large power development project on the Eastmain-Rupert river system in exchange for a cash settlement.

In 2000, a hydroelectric power project that would have flooded an important wetland was rejected. In 2001,

Iceland's National Planning Agency rejected plans for a hydroelectric power project that would have dammed two of the three main rivers flowing from Europe's largest glacier and destroyed a vast wilderness. Since the 1970s, surface air temperatures appear to have increased on average 1.5°C per decade over continental Siberia and western portions of North America, both of which are major sources of freshwater into the Arctic basin. The opposite trend is occurring in Greenland and Canada's eastern Arctic where there is a negative trend of -1°C per decade.

Arctic countries have partially responded to threats to their freshwater systems by establishing protected areas and designating important wetland areas under the Convention on Wetlands of International Importance. Nearly half the protected area in the Arctic is the Greenland ice cap and glaciers which store freshwater. Although the Antarctic ice cap is the world's largest body of freshwater, there are also seasonal streams and rivers, and numerous lakes and ponds in Antarctica.

Other sources of freshwater are captured in the glaciers that occur in many coastal Antarctic regions. All these freshwater features are potentially threatened by pollution, including contaminates introduced by Antarctic scientists and tourists. Freshwater lakes are found mainly in the coastal regions, on many of the sub-Antarctic Islands, and in the rare ice-free areas. Many are exposed to potential contamination from human activities. Observations at selected lakes, however, show that contamination resulting from research activities and station operations is generally near or below detection levels.

Bibliography

ACIA, 2001, Arctic Climate Impact Assessment.

Amap, 1997, *Arctic Pollution Issues: a State of the Arctic Environment Report.* Oslo, Arctic Monitoring and Assessment Programme.

AMAP, 1997, *Arctic Pollution Issues: A State of the Arctic Environment Report.* Oslo, Arctic Monitoring and Assessment Programme.

Caff 2001, *Arctic Flora and Fauna: Status and Conservation.* Helsinki, Arctic Council Programme for the Conservation of Arctic Flora and Fauna.

Crane, K. and Galasso, J.L., 1999, *Arctic Environmental Atlas.* Washington DC, Office of Naval Research, Naval Research Laboratory

EPPR 1997, *Environmental Risk Analysis of Arctic Activities. Risk Analysis Report No. 2.* The Emergency Prevention Preparedness and Response Working Group of the Arctic Council.

Farman, J.C., Gardiner, B.J. and Shanklin, J.D., 1985, *Large losses of total ozone in Antarctica reveals seasonal ClOX/NOZ interaction.* Nature 315, 207-10

Interagency Commission 1998, *The Second National Communication to the UNFCCC.* Moscow, Interagency Commission of the Russian Federation on Climate Change Problems

Interagency Commission, 1998, *The Second National Communication to the UNFCCC.* Moscow, *Interagency Commission of the Russian Federation on Climate Change Problems.*

IPCC, 2001a, *Climate Change 2001: Impacts, Adaptation and Vulnerability.* Contribution of Working Group II to the Third Assessment Report of the Intergovernmental Panel on Climate Change. Cambridge, United Kingdom, and New York, United States, Cambridge University Press

IPCC, 2001a, *Climate Change 2001: The Scientific Basis. Contribution of Working Group I to the Third Assessment Report of the Intergovernmental Panel on Climate Change.* Cambridge, United Kingdom, and New York, United States, Cambridge University Press.

IPCC, 2001b, *Climate Change 2001: Impacts, Adaptation and Vulnerability. Contribution of Working Group II to the Third Assessment Report of the Intergovernmental Panel on Climate Change.* Cambridge, United Kingdom, and New York, United States, Cambridge University Press

IPCC, 2001b, *Climate Change 2001: The Scientific Basis.* Contribution of Working Group I to the Third Assessment Report of the Intergovernmental Panel on Climate Change. Cambridge, United Kingdom, and New York, United States, Cambridge University Press

Kriner, S. 2001b, *Flood Disaster Averted Again in Siberian City.* American Red Cross, 23 May 2001.

NASA, 2001, Largest-ever ozone hole observed over Antarctica. NASA Goddard Space Flight Center.

Neff, W.D., 1999, Decadal time scale trends and variability in the tropospheric circulation over the South Pole. *Journal of Geophysical Research-Atmospheres,* 104, 27217-51

Noaa, 2001, *The Exxon Valdez Oil Spill.* Office of Response and Restoration, National Ocean Service, National Oceanic and Atmospheric Administration.

Index